Social Biology of the Bushy-tailed Woodrat,

Neotoma cinerea

by Peter C. Escherich

A contribution from the
Museum of Vertebrate Zoology,
University of California, Berkeley

UNIVERSITY OF CALIFORNIA PRESS
Berkeley • Los Angeles • London

UNIVERSITY OF CALIFORNIA PUBLICATIONS IN ZOOLOGY

Volume 110

Issue Date: December 1981

UNIVERSITY OF CALIFORNIA PRESS
BERKELEY AND LOS ANGELES, CALIFORNIA

UNIVERSITY OF CALIFORNIA PRESS, LTD.
LONDON, ENGLAND

ISBN 0-520-09647-9
LIBRARY OF CONGRESS CARD CATALOG NUMBER: 81-11492

Library of Congress Cataloging in Publication Data

Escherich, Peter C.
Social biology of the bushy-tailed woodrat,
Neotoma cinerea.

(University of California publications in
zoology; v. 110)
"A contribution from the Museum of Vertebrate
Zoology, University of California, Berkeley."
Bibliography: p.
1. Bushy-tailed wood rat—Behavior. 2. Social
behavior in animals. 3. Mammals—Behavior.
I. Title. II. Series. II. University of Califor-
nia, Berkeley. Museum of Vertebrate Zoology.
III. Series.
QL737.R638E79 599.32'33 81-11492
ISBN 0-520-09647-9 AACR2

Contents

List of Figures

List of Tables

List of Plates

Acknowledgments

I thank George W. Barlow for his support, advice, and encouragement throughout this study as well as his helpful suggestions for the manuscript. William Z. Lidicker, Jr., and Stephen E. Glickman gave useful advice during the study and for improvement of the manuscript. Elmer C. Birney and Dennis Rainey provided helpful critiques before publication. James L. Patton first introduced me to bushy-tailed woodrats and the questions they presented; his advice and encouragement contributed greatly. H. J. Egoscue and J. H. Brown willingly shared unpublished observations of *Neotoma cinerea* from their separate extensive research experiences with the species. K. F. Murray, B. C. Nelson, and C. R. Smith of the Vector Control Section, California Department of Health, provided information concerning *N. cinerea* from their continuing surveys of plague throughout California.

Many other colleagues aided through help in the field, by caring for animals while I was in the field, as well as in stimulating discussion: B. Csuti, W. Glanz, S. Holbrook, F. McCollum, J. Mascarello, G. N. Cameron, C. Nicoll, and others. Discussions with K. Wallen have been especially helpful, for his study of social behavior in *N. fuscipes* was contemporaneous with the present one. Permission to work at Sagehen Creek Field Station, as well as aid in many ways while there, came from A. S. Leopold, M. White, and V. Hawthorne. Space at the Field Station for Animal Behavior Research, on the Berkeley campus, was provided by F. Beach and S. Glickman. Access to the collection of the Museum of Vertebrate Zoology was provided by W. Z. Lidicker, Jr., and J. L. Patton. C. Jones allowed me to study specimens and field notes in the U.S. National Museum, Washington, D.C. S. McGinnis and B. Wheeler gave advice for construction of radiotracking equipment. E. Reid drew base and habitat maps for individual study sites, and she and G. Christman provided advice and materials for preparation of the remaining figures.

Support for this study, in the form of equipment, transportation, and animal care facilities, came from the Department of Zoology and the Museum of Vertebrate Zoology, University of California, Berkeley. Mrs. Frances Escherich typed the original draft, and with my father, Alfred C. Escherich, provided funds for additional typing. Finally, my wife Susan helped in many ways in the field and laboratory, and provided continuing support and encouragement.

Abstract

The social biology of *Neotoma cinerea* was examined in the field, laboratory, and museum. It is the most boreal woodrat and occurs in harems. It shows strong sexual dimorphism and well-developed scent marking.

Over 1000 museum specimens were analyzed in series and by sex and age classes. Sex ratios were equal. Adult and subadult males occurred solitarily more commonly than did females. This difference first appeared at dispersal age, 2 to 3 months. *N. cinerea* is more sexually dimorphic in size than three other *Neotoma* species. Of three *N. cinerea* subspecies, the most dimorphic is *N. c. alticola*, from the most severe climate, which was the subspecies studied in the field.

A laboratory population was maintained and bred for behavioral observations. Some individuals became tame, but most were fearful and inhibited. Kinesthetic memory is important in learning escape routes. Commonly used paths are marked with urine. Feces are reingested, a behavior typical of foliovorous small mammals. Several auditory signals occur.

Males deposit a musky secretion from their well-developed ventral skin gland. Calcareous deposits from urine of all classes build up on rocks in the field, and lab animals duplicate these. An anal gland, better developed in males, produces an apparently odorless secretion. Males have preputial glands.

Males fight vigorously, scent-marking during the fighting. They respond to the odor of other males with threats. Females are less aggressive.

Bushy-tails bred in the laboratory from February through July. Mating behavior includes considerable chasing and olfactory communication. Copulation occurs away from the house. There is no copulatory lock. The young play from three weeks to two months of age. From six weeks, playing grades into fighting with both siblings and parents.

The field study, at Sagehen Creek in the Sierra Nevada, gave information on social organization. At Sagehen, bushy-tails occurred in isolated rock outcrops. Old urine deposits identified potential study sites. Fresh urine marks indicated population levels. Most data came from live-trapping, supplemented with radiotracking.

Breeding females were caught most frequently, but 85 percent of adults of both sexes were captured on the first night of a trapping period. Growth curves allowed me to estimate ages of the young.

Harems were common, with one adult male to 1-3 adult females. This arrangement first appeared in groups of immigrating subadults, in the year preceding breeding. Later immigrants were excluded by earlier arrivals. Young males left home areas by two and a half months, but many young females remained home to breed there the next year. Within rock areas, home ranges overlapped; most individuals occupied separate dens. Outside the rocks, foraging areas of adult females were more exclusive, but overlapped those of the young and the adult male.

Births occurred from April to August at Sagehen. All breeding females had first litters by June 1, and 2 to 3 litters per season. Most young settling in unoccupied areas suitable for winter survival were born by June 1, while most immigrants to poorer areas were born later.

Only 19-38 percent of young N. c. alticola survive to adulthood (field and museum data). Two other subspecies, from milder climates, have 50 percent survival (museum data).

The social organization of N. cinerea reflects the requirement of rocks for winter shelter. To breed, females must share a limited habitat, and in so doing frequently share with related females. Reduction in fitness due to sharing is minimal, but the penalty for not sharing may be failure to breed. These limited rock outcrops are sufficiently small for one male to exclude others. The sexual dimorphism and scent marking result from sexual selection in the harem situation.

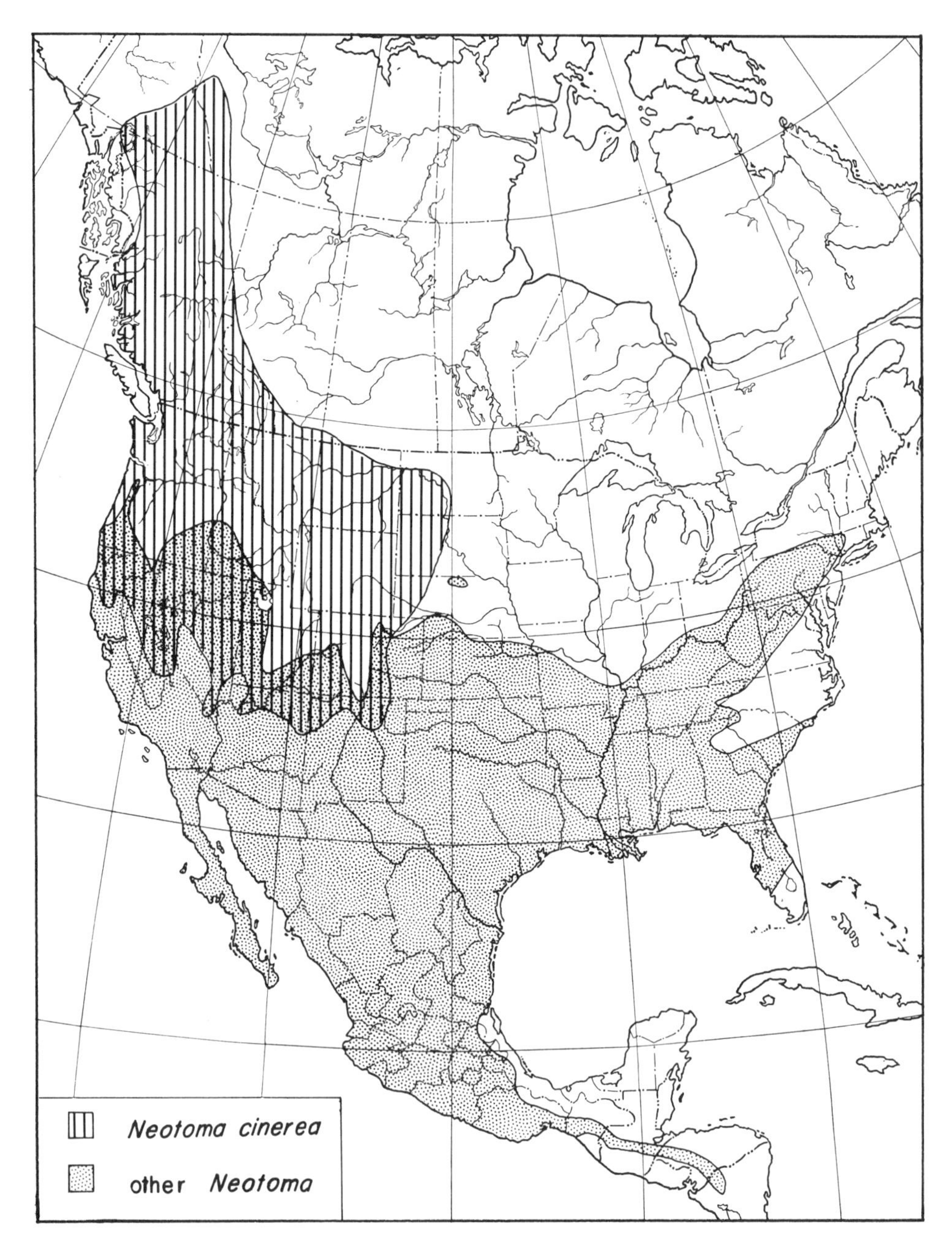

Figure 1. Comparison of the distributions of Neotoma cinerea and other species of Neotoma. (Compiled from Hall and Kelson, 1959, Murray and Barnes, 1968, Birney, 1973, and Martell and Jasper, 1974.)

INTRODUCTION

The present study of the bushy-tailed woodrat, *Neotoma cinerea*, started with an interest in its scent marking. Males rub a musky secretion from their extensive ventral dermal gland (plate 4b) onto objects around them (Egoscue, 1962). In addition, rocky areas where members of the species live can be identified by extensive white urine deposits on the tops and edges of rocks (plates 1a, 5). Although other species of woodrat show some development of a similar ventral gland, in no other is it so well developed. To learn the reasons for such extensively developed scent marking, I needed to investigate *N. cinerea*'s life history and social structure. I discovered woodrats in the field living in harems. This arrangement appears to be rare among nocturnal rodents (see following section), and is probably related to the strong development of scent marking. The main emphasis of this study, therefore, concerned the social organization of this species and various aspects of the species' biology related to it.

The bushy-tailed woodrat lives in high mountain areas of western North America from Colorado to California, and New Mexico to Yukon Territory (fig. 1). Within California, the limits of its distribution in the northern Coastal Ranges and Sierra Nevada can be approximated by the lower level of persistent winter snow (see fig. 2). Such a distribution is unique for a woodrat. Other members of the genus occur either further south (to Central America)(fig. 1) or at lower elevations, typically living in arid shrublands such as chaparral, desert woodland, or tropical scrub.

The most obvious adaptations of *N. cinerea* to its boreal habitat are its bushy tail, thick fur, and large size. An adult male of *N. c. alticola* may weigh from 300 to nearly 600 grams, a female from 250 to 350

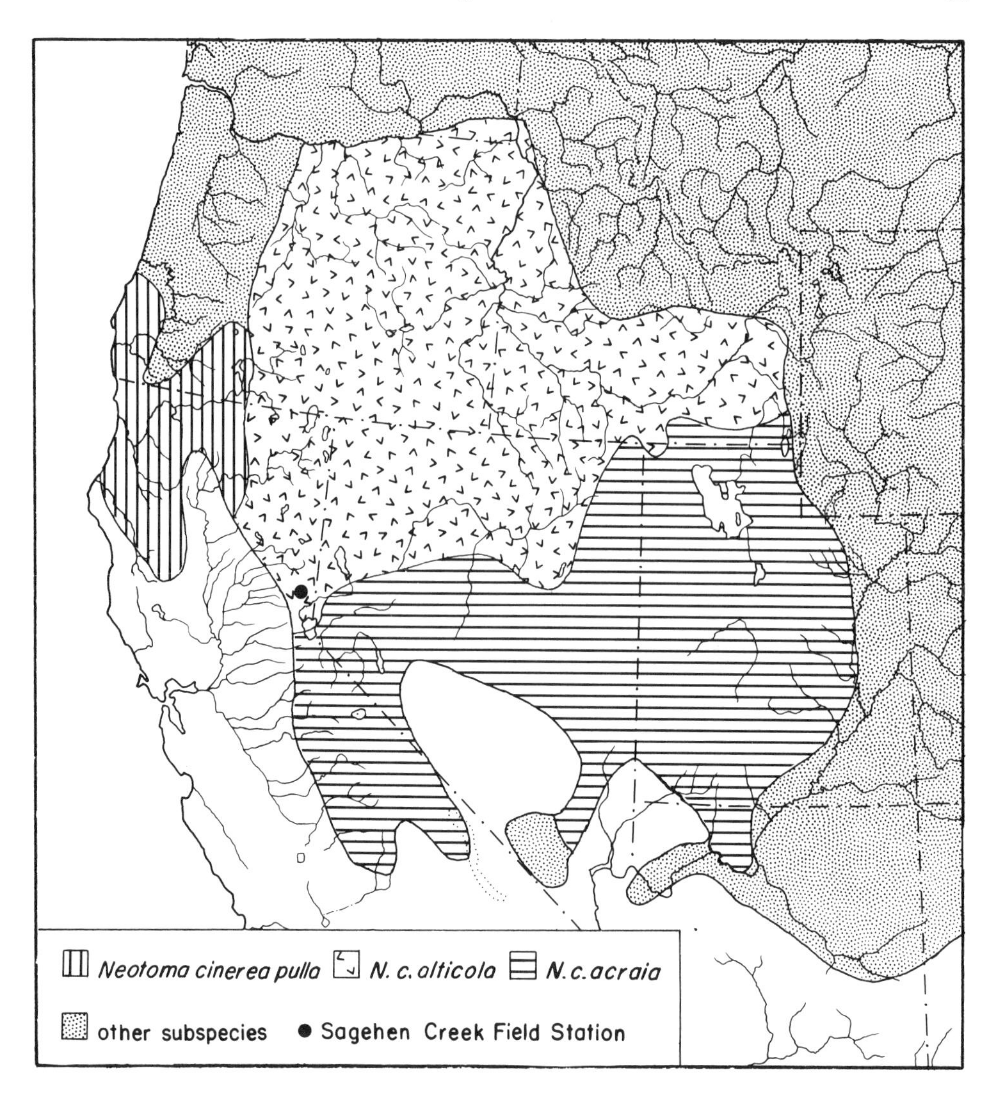

Figure 2. Distribution of three "California" subspecies of Neotoma cinerea. (After Hall and Kelson, 1959, and Hooper, 1940.) Sagehen Creek (solid dot) was location of field study.

grams (table 6). In other species of woodrats, males only average 215-300 grams and females 180-240 grams (table 7).

NOCTURNAL RODENTS AND SOCIOBIOLOGY

Recently there has been increasing interest in questions of the evolutionary basis for social organization. E. O. Wilson's (1975) Sociobiology represents the most comprehensive review; in it (p. 456) he states that "...polygyny is the rule in mammalian systems and harem formation is common." As Wilson says, the reduced role of the male in parental care, due to the dependence of young on the mother's milk, would tend to promote such a social structure. On the other hand, Emlen and Oring (1977) point out that differential parental care in mammals does not require polygyny, but rather provides the potential for it. Whether that potential is realized depends on other factors such as the distribution of resources. They suggest that monogamy may occur more often in mammals than is generally recognized.

A closer examination of mammalian literature finds that practically all detailed field studies of mammalian social structure have concerned large, diurnal, group-living mammals, which live in relatively open habitats. These are mainly ungulates, primates, larger carnivores (including pinnipeds), and ground dwelling squirrels (including marmots and prairie dogs). The reasons for this bias in choice of study animals are obvious: such animals can be seen easily, there are more in one place, and return of data for time invested can be much greater.

The great majority of mammals, however, are small nocturnal forms, as were the ancestors of all present mammals. Nearly 73 percent of all mammalian species are rodents, bats, or insectivores (species counts from Vaughan, 1972). As Eisenberg (1966) points out, adaptive strategy for the smallest mammals differs from that of the largest, particularly with regard to predation. Small nocturnal mammals, especially in habitats with good cover, use secretiveness to avoid predation; only rarely should such mammals form large social groupings. Diurnality, large size, and open habitats not only allow but may select for group living (cf. W. D. Hamilton, 1971). (I greatly oversimplify Eisenberg's arguments here; his review discusses many other factors contributing to mammalian social structures. Many of his then-speculative points have been borne out by more recent field and experimental evidence.)

While conducting the present study, I also reviewed much of the available information for social systems of North American cricetine rodents (peromyscine-neotomine rodents of Hooper, 1968), the group which includes Neotoma as well as Peromyscus (deer mice), Onychomys (grasshopper mice), and a few other genera. This group has received considerable attention from zoologists because of the diverse habitats its members occupy and its proximity to the many investigators in North America. Most of the work has been either laboratory-oriented ethological studies or field studies with purposes other than studying relationships between individuals. As expected, the predominant social systems consist of solitary animals or, at most, pairs and pairs with offspring. With one possible exception, Neotoma lepida (based on preliminary data: Keith Justice, personal communication), no peromyscine-neotomine rodent other than N. cinerea has been demonstrated to live and breed in a polygynous fashion.

BIOLOGY OF NEOTOMA

Woodrats, or pack rats, are well known for two traits: their large houses built of sticks and other debris, and their propensity to pick up and carry away ("pack") almost any loose object. This latter activity has given them a certain notoriety, for many people have fallen victim to it, often with amusing consequences. John Muir, the mountaineer and conservationist, once had to battle a large bushy-tail to keep his ice axe from following his barometer and spectacles into the night (Muir, 1901). Less amused are the owners of summer cabins who may return in spring to find their possessions totally rearranged by a woodrat which found winter shelter in their cabin.

Before proceeding further, definition of a few terms is in order. "House" or "den" refer to the shelter a woodrat constructs from sticks and other material. Rock dwelling woodrat species, such as N. cinerea and N. lepida, may only chink in the openings between rocks (plate 1a); N. fuscipes may build a house 2 m high or more from sticks (plate 1b). "Nest" refers just to the nest cup itself, for occasionally a nest may be found without a house.

Most woodrats are limited within their habitat in the choice of places where they can build houses or find sufficient shelter to build nests. For N. cinerea this usually means suitable rock outcrops, rock slides, caves, or buildings. Such discrete sites which are apparently suitable for woodrats will be referred to here as woodrat "areas," whether or not they are currently occupied. This habitat limitation leads to concentrations of woodrats in such areas. One is tempted to call these colonies, but this word has acquired some ambiguity through varying usages by biologists (E. O. Wilson, 1975). I prefer a more neutral term, "settlement," a woodrat area with one or more residents. Finally, the word "woodrat" refers generally to the genus Neotoma. When referring to N. cinerea or other particular species I either qualify the word or use the name of the species, unless it is clear from context. N. cinerea is often referred to as "bushy-tail."

Several species of woodrats have had extensive monographs devoted to their ecology. Linsdale and Tevis (1951) followed a large population of N. fuscipes, the dusky-footed woodrat, in California. Their study encompassed ten years, and many aspects of the species' life history were recorded by them and by numerous coworkers. Finley (1958) studied the six woodrat species occurring in Colorado. Since N. cinerea is the most common woodrat there, much of his monograph deals with that species. He primarily emphasized distribution and factors that influence it such as food, shelter, and habitat; other aspects of natural history and taxonomy were also discussed. Two other species have received lengthy treatment: N. floridana, the eastern woodrat, by Rainey (1956) and Fitch and Rainey (1956), and N. albigula, the white-throated woodrat, by Vorhies and Taylor (1940). So many shorter studies have been carried out on woodrats that a complete review is impractical here.

From these studies a general picture of woodrat biology emerges. Although food preferences may vary among and within species, most have a broad and flexible diet, with local specializations sometimes occurring. Green leaves or cactus pads dominate the diet and may provide the only source of water (Finley, 1958; Lee, 1963; Meserve, 1974; Schmidt-Nielson, 1964; Stones and Hayward, 1968). Fruits, seeds, and fungi provide most of the remaining food, depending on species and habitat. Woodrats may store considerable quantities of food in the house. Houses are almost never constructed in the open; some form of support such as a suitable bush, tree, or rock crevice provides the necessary site (Olsen, 1973). Most workers agree that availability of house sites limits woodrats more than does availability of food (e.g. Box, 1959; J. H. Brown et. al., 1972; Rainey, 1956; Turkowski and Watkins, 1976).

ORGANIZATION OF PAPER

The data are presented here in three sections representing their three sources: museum, laboratory, and field. The field study provides the context in which the museum and laboratory observations are meaningful. It is presented last, both because it brings together the results of the other two sections and because it leads most naturally into the final discussion, which primarily concerns the social organization seen in the field. For a quick overview of the work, the reader may wish to see the first and last parts of the field section and the final discussion before reading the rest of the report in detail.

MUSEUM MATERIALS

Early in the field work, at Sagehen Creek in the Sierra Nevada, I found bushy-tails living in harem situations. Single adult males occurred in single isolated rock outcrops with one to three adult females. The number of woodrat settlements convenient to the field station, however, seemed small for convincing statistics. Although the number of bushy-tails in a particular locality may appear high on occasion, I doubt whether many places in the species' range contain numerous suitable breeding areas much closer together than those near Sagehen Creek. This patchy distribution would have made direct collection of data for a significantly larger sample difficult. This distribution, however, made it possible to use the extensive museum collections of *Neotoma cinerea*, and accompanying field notes, for the same purpose.

Specimens examined were from the Museum of Vertebrate Zoology (MVZ), University of California, Berkeley, and the Biological Survey Collection of the U.S. National Museum (USNM). Most were collected between 1895 and 1920 (USNM) or 1910 to 1940 (MVZ) in systematic surveys of the western United States for establishing the distributions of mammal species.

The normal technique was for a collector to travel through an area, staying for one to several nights at a given locality, and at each locality setting traplines through a variety of habitats to catch a diverse sample of the local mammal fauna. In so doing, he was unlikely to cover more than one bushy-tail settlement or area at a time, since in most places these are well separated. Thus, in analyzing the museum specimens, a single series (a group of specimens labeled with the same locality and collected within a few days of each other) generally represents the inhabitants of a single rock outcrop, slide, or other suitable habitat area.

Where an unusually large number of individual specimens made up a single "series," the collector's field notes often indicated how many actual sites were represented, and indicated where each specimen came from. In those cases, where appropriate and possible, I designated subseries for purposes of analysis. MVZ collectors kept daily records of their collections, which proved more useful than the weekly summaries of the earlier Biological Survey collectors. Where assignment to subseries was questionable, the conservative approach of leaving individuals together as coming from a single settlement was taken.

METHODS

Over 1000 specimens of *N. cinerea* were examined. Table 1 gives distribution by subspecies and museum. For each individual specimen the following data from the label were recorded: museum number, locality, date, sex, total length, and tail length. Weight and reproductive condition were taken from the label if available. Study skins of lactating or pregnant females could be identified by enlarged nipples surrounded by bare skin. Each individual's age was estimated according to criteria given below. For large series, collectors' names and field numbers were recorded for later reference to field notes. A few specimens were eliminated from the analysis because of insufficient data, usually because the collector failed to record the sex or because a missing skull precluded satisfactory aging.

For the three subspecies which occur in California, *N. c. alticola*, *N. c. acraia*, and *N. c. pulla* (Hall and Kelson, 1959, pp. 701-705), all

TABLE 1
Museum Specimens of *Neotoma cinerea* Examined

Subspecies	Museum		Totals
	MVZ	USNM	
N. c. alticola	198	265	463
N. c. acraia	189	78	267
N. c. pulla	80	59	139
Other (groups only)	115	34	149
Totals	582	436	1018

specimens from both museums were examined for the entire range of the subspecies, in each case extending beyond California (fig. 2). MVZ contained considerably fewer specimens of subspecies other than these three, so for other subspecies only series which contained more than one specimen ("group") were examined; single specimens were ignored. These latter series were analyzed separately.

Aging Procedure

Finley (1958, pp. 231-240) provided detailed descriptions of molt stages, dentition changes, and other features useful in estimating age of *N. cinerea*. Egoscue (1962) described the postjuvenal molt, adding to Finley's descriptions. In general, I have followed the age classes defined by Finley, although with certain modifications, described below, which take into account more recent knowledge of growth and reproductive biology of the species (Egoscue, 1962; Martin, 1973; present study). Skull and pelage characters given here are the primary ones used in aging specimens for the present study; see Finley (1958) for more details of pelage and skull development.

Juveniles

Nestlings up to three weeks of age. Growth curves (Egoscue, 1962; Martin, 1973; present study) show that animals may reach 75 grams within that time, not 50 grams as described by Finley for Colorado animals. First and second molars have erupted, but show little or no wear at three weeks. Dorsal pelage black, changing to gray, the tail heavily furred at three weeks, but not bushy as in the adult (plate 2a). Juveniles are relatively uncommon in museum collections because they remain in the house during this period and are not trapped. Those in collections generally result from a collector dismantling a house or finding a nest in a cave or building.

Immatures

Three weeks to two months; weight 75-175 grams. Young are weaned between three and four weeks of age and may disperse soon after weaning, although more commonly they remain in the house area until reaching at least the subadult stage (see field section). Juvenal pelage gray, tail well furred, but not bushy (plate 2b and c). Third molar erupted, but showing little or no wear. At about six weeks of age growth slows

(Egoscue, 1962; present study, figs. 6 and 7), and the postjuvenal molt starts.

Subadults

It is in distinguishing this stage from the adult that I differ most from Finley. Bushy-tails are seasonally polyestrous and do not breed the year of their birth (Egoscue, 1962; present study). In late summer in the field I occasionally found an older first year female with an open vagina, but none was pregnant. Thus I prefer to consider all animals subadult or younger until the beginning of their first breeding season. Finley (1958) considered *N. cinerea* adult at four months.

I arbitrarily separated subadult from the adult class at January 1 following birth because the breeding season may start in February or March, and virtually no bushy-tails have been collected in winter, minimizing the number of specimens to be separated in this way.

An older subadult may be nearly indistinguishable from an adult using size and pelage alone (plate 2d; cf. plate 4a and c). A few individuals, as noted by Finley, retain a patch of grayish juvenal pelage on the head long after completing the rest of their postjuvenal molt. Finley described a whitish tail tip in animals of this age; it was seen only rarely in the subspecies examined here.

The upper molar teeth, on the other hand, provide a reliable index. Viewed from the sides, the folds (reentrant angles) of subadult molars extend well into the alveoli (plate 3a). As molars grow during the first fall and winter, upper limits of the folds emerge from the bone, so that in adults a smooth lateral surface can be seen on the molar above the folds, and part of the root appears (plate 3b).

Of all specimens studied, no female designated subadult by this criterion showed the bare and enlarged nipples typical of breeding, confirming the aging technique. Since birth takes place from March to August, transition to adulthood (here arbitrarily January 1) occurs between five and ten months of age.

Adults

The criteria for separating adults and subadults have already been discussed. Beyond initial growth of the molars, I considered tooth wear too variable to separate year classes of adults reliably. Also, adults can breed in more than one year if they survive, so further separation was unnecessary for present purposes.

Finley distinguished a "senile" class which I did not use. The oldest animals may have the molar folds completely worn away and much of the roots exposed. The last adult molt may be skipped, resulting in a worn brownish pelage because of missing black guard hairs. In the laboratory, three individuals survived beyond five years, and eleven others lived between three and five years. The oldest known individual in the field was a male (#1598) first captured as an adult in August 1971, and last seen in July 1973, suggesting a maximum recorded field longevity of three years. Two females (#1134, 1588) were known to have lived at least two years in the field. Finley had suggested a maximum age of two and one half years.

ANALYSIS

The data were treated in a number of ways to test the hypothesis of polygynous social structure. The first step involved counting the number of individuals of each age class and sex in each series as well as the total number of individuals per series. Because there were few juveniles or immatures, these two categories were combined. The definition of "series" given previously is somewhat imprecise because in many cases the final decision on what to include in a series depended on careful evaluation of collectors' field notes. The aim was to represent a single woodrat settlement by each series or subseries. This ideal could not always be realized, and many series included specimens that might have been separated into subseries given adequate field notes. This approach leads to an underestimate of the degree of polygyny and makes any conclusions reached concerning that question conservative.

Because of the possibility of climatic differences differentially affecting social biology of different subspecies, analyses were carried out not only collectively for all specimens recorded, but also separately for each of the three subspecies whose distributions include California (see fig. 2). These separate analyses are reported where they proved of some interest.

Differential trappability of different age and sex classes could have biased results. The field population at Sagehen Creek was well known as a result of continued long-term live-trapping. Examination of those data (see field section for details) indicated that adult females were recaptured more frequently than other classes, but that there was no difference between the sexes in the proportion of those known present which were captured on the first night of live-trapping. Because recapture rates have little relevance to animals kill-trapped for museum collections, the museum samples should represent the actual ratios of sex and age classes.

Sex Ratio

If apparent polygyny is seen in a population, one of the first questions is whether it is due to (or possibly causes) an unbalanced sex ratio. A corollary is whether any change occurs in sex ratio from younger to older age classes.

To answer these questions, only data for the three "California" subspecies were used, because these data sets included all animals collected. Use of the other subspecies' data here could have biased the results, since only animals occurring in groups were examined. Table 2 shows the total number of individuals of each class summarized, and separately by subspecies, for _N. c. acraia_, _N. c. alticola_, and _N. c. pulla_. Neither the summarized data nor those for individual subspecies are significantly different from an even ratio (Chi-square values given in table), indicating no change in sex ratio with age.

Two methods were used to test the sex ratio. First, the three subspecies, each divided into three age classes, constitute nine paired samples: six have more females, two have more males, and one is equal. This distribution is not different from chance (one-tailed sign test; p = 0.145). Second, the two most extreme sex differences in the table were tested against the hypothesis of equality. Adults of _N. c. pulla_ showed a ratio of 33 males to 41 females (M/F = 0.80); there were 8 males to 12 female juvenile-immatures (M/F = 0.67) of that subspecies. Neither of these ratios is significantly different from 1 (binomial test; p = 0.21 and 0.25, respectively).

Thus by neither method of testing could any difference from equality be shown in the sex ratio.

TABLE 2
Sex Ratio and Age Class: Individuals of Three "California" Subspecies of *Neotoma cinerea*
(Specimens from MVZ and USNM; see text for explanation)

Subspecies Sex	Age Class[a] Adult	SbAd	Jv-Im	Totals	Chi-square
a. Three subspecies combined					
Males	183	186	56	425	
Females	199	187	58	444	
Totals	382	373	114	869	0.292, n.s.
b. *N. c. acraia* alone					
Males	61	46	21	128	
Females	69	47	23	139	
Totals	130	93	44	267	0.141, n.s.
c. *N. c. alticola* alone					
Males	89	116	27	232	
Females	89	119	23	231	
Totals	178	235	50	463	0.356, n.s.
d. *N. c. pulla* alone					
Males	33	24	8	65	
Females	41	21	12	74	
Totals	74	45	20	139	1.287, n.s.

a. SbAd: Subadult
Jv-Im: Juvenile-Immature

Group Structure

Because of the difficulty in separating museum series into subseries which I felt more accurately represented single woodrat settlements, analysis of the polygynous structure was not so simple as originally hoped, and a number of indirect methods had to be used.

Since the sex ratio is apparently even, a polygynous structure should lead to an excess of adult males living by themselves, and conversely, most adult females should be living with other conspecifics (my own field work suggested this to be true). The first method of analysis was based on this assumption. For each sex and age class of the "California" subspecies, the individuals which had been collected in a series with other bushy-tails ("group") were counted and compared with the number of that class collected singly. The sex and age of other individuals in the series were ignored. Therefore, individuals labeled "group" here were not necessarily collected as members of breeding groups nor were they necessarily living with other members of their class, but were simply collected near other conspecifics. This method measures the relative tendency of members of a particular class to occur solitarily, by identifying which classes were more frequently collected at sites where no other bushy-tails were collected. Based on my field experiences and those of other workers (field notes, publications, and personal communications), these "solitary" individuals tend to be those occurring in marginal habitat: often smaller rock outcrops or other areas with little good shelter available.

The results of this analysis are summarized in table 3. Of 869 animals represented, 693 (80%) were collected with at least one other conspecific. Only 176 (20%) were collected as solitary individuals, demonstrating the degree to which these animals generally tend to live near each other.

TABLE 3
Individuals of *Neotoma cinerea* Occurring Solitarily and in Groups: Comparisons Within Age Classes
(Specimens from MVZ and USNM; see text for explanation)

Age Class Sex	Occurrence In Group	 Single	Totals	% Single	Chi-square (cont.corr.)
a. Adult					
Males	140	43	183	23%	
Females	172	27	199	14%	
Combined	312	70	382	18%	5.63, $p<0.02$
b. Subadult					
Males	132	54	186	29%	
Females	151	36	187	19%	
Combined	283	90	373	24%	4.35, $p<0.05$
c. Juvenile-Immature					
Males	50	6	56	11%	
Females	48	10	58	17%	
Combined	98	16	114	14%	0.54, n.s.

As predicted, adult males do occur solitarily more commonly than do females (table 3a). Subadults, although not yet capable of breeding, also show this difference (table 3b), indicating that the polygynous structure is set up at the time of dispersal (field data from my mark-recapture studies support this observation). The juvenile-immatures, most of which would have been collected where they were born, do not show this tendency (table 3c).

Table 3 suggests a number of other comparisons, particularly regarding the differences between age classes. First, the fraction of subadults found alone appears larger than the fraction of solitary adults. When all adults are compared with all subadults (table 4a), the differ-

TABLE 4
Individuals of Neotoma cinerea Occurring Solitarily and in Groups: Selected Comparisons between Age Classes (Specimens from MVZ and USNM; see text for explanation)

Comparison / Age Class[a]	Occurrence: In Group	Occurrence: Single	Totals	% Single	Chi-Square (cont. corr.)
a. All post-immatures					
Adults	312	70	382	18%	
Subadults	283	90	373	24%	
Combined	595	160	755	21%	3.47, $p<0.1$
b. First-year males					
Subadults	132	54	186	29%	
Jv-Im	50	6	56	11%	
Combined	182	60	242	25%	6.79, $p<0.02$
c. All males					
Ad plus SbAd	272	97	369	26%	
Jv-Im	50	6	56	11%	
Combined	322	103	425	24%	5.60, $p<0.02$
d. All females					
Ad plus SbAd	323	63	386	16%	
Jv-Im	48	10	58	17%	
Combined	371	73	444	16%	0.00019, n.s.

a. Ad: Adult
SbAd: Subadult
Jv-Im: Juvenile-Immature

ence is found to be only weakly significant ($p<0.1$). If the same comparison is made within sexes (not shown), the differences are also not significant.

As noted, juvenile-immature individuals do not show the degree of solitariness seen in older age classes. The origin of this difference can be seen by comparing the youngest class against the older classes within each sex. Juvenile-immature males show a significantly different distribution both from subadult males and from all older males (table 4b and c). On the other hand (table 4d) the distribution of females is virtually identical between the younger and older age classes. These differences suggest that the polygynous structure may be controlled by males excluding other males more strongly as they mature, while females show little change.

To provide a somewhat independent test of the presence of polygyny in the museum "population," as well as extending the tests to other subspecies, another approach was used. With the intention of more closely sampling actual breeding groups, only series which included at least one adult male and one adult female were counted. These series were classified as to whether they contained more adult males, equal numbers of adult males and females, or more adult females (see table 5). Because of the nature of the original collections, in which incomplete collections of several settlements might be included in a single series, and my own inability to properly separate these, a certain number of series would be expected to contain excess males. In a polygynous system, however, the number of series containing excess females should exceed those

TABLE 5
Relative Numbers of Males and Females in Museum Series
of *Neotoma cinerea* Containing at Least One Adult of Each Sex
(All subspecies examined)

Ratios of Adults	Subspecies: *N. c. acraia*	*N. c. alticola*	*N. c. pulla*	Other Subsp.	Totals
More males	2	4	3	1	10[a]
Males and females equal	12	8	4	5	29
More females	6	8	5	5	24[a]

a. Binomial test, 10 vs. 24; $z = -2.229$; $p = .013$

containing excess males. The "equal" category would contain monogamous pairs and multiple samples of these as well as incomplete samples of either of the other categories.

Presence of many apparently monogamous situations in this system is not inconsistent with the emphasis given here to polygyny. In my field work I found no cases of a female breeding in an area where she had not been present by at least the preceding August. Males, on the other hand, were not uncommonly replaced between breeding seasons, but none were replaced during the breeding season. Thus, from a potentially polygynous arrangement of subadults and adults in the fall, losses of individuals would have different outcomes for the two sexes. A lost male may be replaced from the pool of solitary males; a lost female is not replaced, and the result may tend toward a more even sex ratio within the breeding group.

A significantly greater number of series contained more females than males, thus supporting the original hypothesis of polygyny. Each separate subsample shows the same trend, although sample sizes are too small to produce persuasive statistics separately.

Sexual Dimorphism

Sexual dimorphism is commonly associated with polygamy (Darwin, 1871; reviews in Campbell, 1972). Data from the museum sample were used to compare the degree of dimorphism between adults of the three "California" subspecies of N. cinerea, as well as providing data for comparison with values calculated for other woodrats from published data.

Two measurements were used in assessing dimorphism. The weight, as recorded in the field by the collector, provided the first. The second, head and body length, was calculated from recorded total length minus recorded tail length. This second measurement was used because weights were not always recorded, and because it should provide a more constant index of overall body size somewhat independent of condition, pregnancy, and other factors.

The ventral dermal gland is an obviously sexually dimorphic feature. It covers much of the ventral surface of mature males (plate 4b), and is missing or poorly developed in females. It was not considered suitable for the present analysis, however, because many factors besides sex would have reduced reliability and made it a less useful measure of dimorphism than length and weight. These problems include: unreliability of measurements taken from dry study skins (besides stretching when skinned and stuffed, and shrinking when dried, the belly skin is cut and sewed in preparing the specimen); seasonal and molt differences; effect of different colored soils (in which the live animal had dustbathed) on apparent extent of the gland. Even without detailed analysis, comparison between study skins of *N. cinerea* and other species shows that the ventral gland is considerably more extensive in males of the former, and restricted to a narrow strip down the midline in other species (see also Reynolds, 1966; Howell, 1926). Apparent differences in the ventral gland between subspecies are discussed in the laboratory section of the present report.

N. cinerea: Within-Species Variation

Samples used here were considerably smaller than those of previous sections because I wanted to use both weight and length measurements taken from the same specimens. Biological Survey collectors did not record weights, so I used only MVZ specimens, many of which were weighed by the collector in the field. I eliminated many females which collectors indicated were pregnant, but could not eliminate all such because reproductive condition was not always recorded by collectors.

For each sex and subspecies, the mean, range, standard deviation, and sample size are given for weight and for head and body length in table 6. For each subspecies, the ratio of female to male weight and of female to male length were calculated. To find a value more nearly comparable to the weight ratio, I also calculated the cubes of the length ratios, which I designated the "$\overline{L}^3$ ratio." If we assume that woodrats are of a relatively constant shape, the $\overline{L}^3$ ratio and weight ratio should be similar within each sample. A consistency index was calculated by dividing the weight ratio by the $\overline{L}^3$ ratio; values of this index close to 1.0 are considered consistent, and values deviating from 1.0 (higher or lower) are less consistent. Finally, the mean was taken for the two ratios to provide a combined dimorphism index.

Of the three subspecies, the most dimorphic is *N. c. alticola*, which is also the largest and most northern. The least dimorphic is *N. c. pulla*, the smallest subspecies. Rats of this taxon live in the Coast Ranges of California and Oregon, and experience a milder climate than do

TABLE 6

Weight and Length Dimorphism in 3 Subspecies of *Neotoma cinerea*
(Specimens from Museum of Vertebrate Zoology)

Subspecies	Sex	Weight (grams)					Head and body length (cm)						Consistency index[b]	Mean dimorphism[c]
		$\bar{W}$	Range	SD	N	$\bar{W}$ ratio[a]	$\bar{L}$	Range	SD	N	$\bar{L}$ ratio[a]	$\bar{L}^3$ ratio[a]		
N. c. acraia	F	250	175–348	49.1	16	.797	202	175–223	12.1	16	.939	.822	.970	.810
	M	313	181–459	96.4	8		216	195–234	14.4	8				
N. c. alticola	F	308	257–366	29.9	20	.724	215	185–233	11.8	20	.906	.743	.974	.734
	M	425	283–585	82.8	18		238	214–265	15.9	18				
N. c. pulla	F	243	166–361	51.2	18	.923	199	175–217	11.6	18	.969	.910	1.014	.917
	M	263	192–322	38.6	11		205	189–215	8.0	11				

a. Female/Male
b. Consistency index = $\bar{W}$ ratio/$\bar{L}^3$ ratio
c. Mean dimorphism = ($\bar{W}$ ratio + $\bar{L}^3$ ratio)/2

TABLE 7

Weight and Length Dimorphism in 3 Species of *Neotoma* for Comparison with Data in Table 6 Regarding *N. cinerea*
(Original data from literature; see text for references.)

Subspecies	Sex	Weight (grams)					Head and body length (cm)						Consistency index[b]	Mean dimorphism[c]
		$\bar{W}$	Range	SD	N	$\bar{W}$ ratio[a]	$\bar{L}$	Range	SD	N	$\bar{L}$ ratio[a]	$\bar{L}^3$ ratio[a]		
N. albigula	F	181	142–260	—	92	.834	178	(109–192)	—	92	.932	.809	1.031	.822
	M	217	136–294	—	80		191	(187–206)	—	80				
N. floridana	F	216	174–260	—	14	.722	212	(170–227)	—	20	.991	.972	.743	.847
	M	299	220–384	—	21		214	(175–270)	—	28				
N. fuscipes	F	239	184–357	38.1	20	.945	217	205–236	7.4	20	.986	.960	.984	.953
	M	253	205–302	24.8	20		220	200–230	8.3	20				

a. Female/Male
b. Consistency index = $\bar{W}$ ratio/$\bar{L}^3$ ratio
c. Mean dimorphism = ($\bar{W}$ ratio + $\bar{L}^3$ ratio)/2

those of the other two subspecies. Brown and Lee (1969) pointed out that woodrats living in more severe climates are larger and better adapted physiologically to cold. Since the dimorphism follows a similar trend, it would suggest that sexual dimorphism in this species is also related to more severe climates. This point is discussed more fully in the final discussion.

Comparison to Other Woodrat Species

To get a broader perspective on the dimorphism in N. cinerea, values for the degree of dimorphism in other woodrat species were calculated from published data. Linsdale and Tevis (1951, p. 438) supplied a data table of weight and measurements for N. fuscipes; these data could be treated identically to those for N. cinerea. For N. albigula and N. floridana, only means and ranges were available (Vorhies and Taylor, 1940, p. 459; Rainey, 1956, pp. 542-543), so no standard deviations could be calculated. The ranges for head and body length I calculated from their ranges for total and tail length; since different individuals might be represented in the extremes for each sample, the ranges in the table are considered approximate. (Birney, 1973, published further length measurements for N. floridana and N. micropus, but did not include weights for those specimens. Although his figures were not used here, they may provide useful material for further work on geographic variation in dimorphism.)

N. floridana apparently shows a strong degree of weight dimorphism, but several factors suggest that value is not directly comparable to the data for other species. In all other taxa in tables 6 and 7, weights and lengths came from the same animals, and all have consistency indices within about 3 percent of 1.0. Rainey took weights from live-trapped animals in his field study, and lengths from museum specimens; the consistency index differs by more than 25 percent from 1.0, eight times the difference found for any other sample, suggesting that his museum and field animals were dissimilar samples. Rainey excluded all pregnant females from his field sample, which I could not do for my museum samples, and the other workers may not have done so either, although their procedures are not clear from their papers. Excluding even early pregnant females would tend to increase apparent dimorphism not only by removing weight of unborn fetuses, but also by possibly selecting younger, and therefore smaller females. Conversely, by not totally excluding pregnant females in the other samples, their degree of dimorphism is underestimated.

Field Data: N. cinerea alticola

During my field work at Sagehen Creek I regularly weighed live-trapped N. cinerea alticola (see field section for procedures). Although the total number of individuals was smaller than for any of the samples just discussed, analysis of dimorphism in this field population is important for several reasons. Because I was following animals through the season, I could eliminate any weights of pregnant females. This permits direct comparison with Rainey's field data. My field population belongs to the subspecies of N. cinerea that showed the most extreme dimorphism, so the field data provide an independent comparison with the museum sample. Finally, since there are weight changes over the season (animals at Sagehen tended to gain weight in the fall), with multiple weights for the same individual I could bracket the range of variation in sexual dimorphism within the same local population.

I first listed the maximimum (usually the first) weights found during each trapping period for each adult (for females, only when they were not pregnant) since animals usually lost weight after spending one or more nights in a trap. From the weights listed for each animal I selected the maximum and minimum weights found over its adult life, and took averages within each sex of these extreme values (table 8). (No lengths were taken of live animals.) A small number of males became too heavy late in the year for my combination of scale and weighing bag, so those weights had to be listed as "over 450 gm"; in calculating means these were counted as equal to 450 gm, underestimating slightly both the mean maximum weight of males and the degree of dimorphism in ratios using that mean.

At the bottom of table 8 are listed the dimorphism ratios for the possible combinations of average weight. It is interesting, although possibly coincidental, that the ratios of maximum weights and of minimum weights are almost identical. These values are lower (more dimorphic) than any of the ratios in tables 6 or 7, possibly indicating the value of controlling for pregnancy when estimating dimorphism. The other two

TABLE 8
Weight Dimorphism in Live Adult
Neotoma cinerea alticola
(Field data: maxima and minima for the same individuals, females not pregnant)

Sex	Max./Min.	Weight (grams)			N
		$\overline{W}$	Range	SD	
Females	Max.	302	250–330	11.1	9
	Min.	279	230–300	10.8	
Males	Max.	440+	405–450+	28.6	7
	Min.	409	365–445	13.1	

"Uniform" ratios—bias minimized by using all maximum values for both sexes or all minimum values; suggest typical values for weight dimorphism.

$\frac{\overline{W} \text{ max F}}{\overline{W} \text{ max M}}$ = .685, or less $\frac{\overline{W} \text{ min F}}{\overline{W} \text{ min M}}$ = .683

Most biased ratios possible—minimum for one sex, maximum for other; suggest extreme values possible for weight dimorphism.

$\frac{\overline{W} \text{ min F}}{\overline{W} \text{ max M}}$ = .634, or less $\frac{\overline{W} \text{ max F}}{\overline{W} \text{ min M}}$ = .738

ratios were calculated to show the potential range for the dimorphism ratio if each individual had been sampled only once, as was done for all other samples here, and had somehow been selected to produce the most biased and unlikely ratios possible. Even the highest (least dimorphic) ratio is not much higher than that for the museum sample of *N. c. alticola*. The strong degree of dimorphism in the field sample of *N. c. alticola* confirms the above finding that this subspecies, the largest and most boreal taxon examined, is also the most dimorphic. The significance of these data is dealt with further in the final discussion.

LABORATORY OBSERVATIONS

A laboratory colony of bushy-tails was established in Berkeley for behavioral observations; of these, thirty-six were brought in from the field, and thirty-nine more were born in the lab. Particular attention was given to interactions between individuals. These included mating, aggressive, and parental behaviors, as well as scent marking and other potentially communicatory behaviors. Observations on the developing young were helpful in explaining phenomena seen in the field.

To give a broader perspective by understanding differences between this species and other congeners, a small number (4 wild, 18 lab-born) of *Neotoma lepida*, the desert woodrat, were kept and bred. Even fewer (five) *N. fuscipes*, the dusky-footed woodrat, were maintained but not bred.

METHODS

The colony started with three animals trapped in February 1970 near Susanville, Lassen County, California. Later, animals were added from near Sierraville (Nevada County) and sites in Modoc County, California. All were of the subspecies *N. c. alticola*, the same one studied in the field.

General procedures for maintaining the colony followed those used by Egoscue (1962) for *N. c. acraia* in Utah.

Cages of three sizes were used: (1) Single animals were maintained in plastic-pan mouse cages, 47 x 38 cm, containing pine shavings for bedding. (2) "Office" cages of 1.5 cm wire mesh, 0.5 or 1 x 0.6 x 1 m, were used for individuals, pairs, and families. Waste dropped through the mesh floor into trays with pine shavings. Paper and sticks were supplied for nests. (3) Additional individuals, pairs, or families were housed in large observation cages, 2.7 x 2.1 x 2.4 m, at the Animal Behavior Research Station on the Berkeley campus. These cages have cement floors, and one side is open (wire mesh) to the outside. Shelves, wooden boxes nailed to the wall, and plywood sheets leaned against the wall provided places for the rats to store food and build houses from sticks and leaves provided.

The latter two types of cages were the source of most observations cited here. The "office" cages literally surrounded my desk for more than four years and were occupied for most of that time. This arrangement was necessary because the rats were generally nervous and reacted to most disturbances by running to their nests and curling up, often for hours. By having the cages in my office I was able to habituate some animals to my presence, and by working quietly at my desk until the rats became active I was able to observe more than I might have otherwise.

The observation cages at the behavior station provided more space for the animals, a situation more comparable to the field than the office cages. On the other hand, the animals habituated poorly and were more fearful in the larger cages, and I was less able to observe their behavior, except for tamer individuals, arranged encounters, and occasional baiting with fresh food. Direct observation in the field was almost completely precluded because houses, food stores, and most activity occurred under several layers of large boulders or in heavy brush.

Laboratory mouse chow was available in excess. Usually rabbit chow or horse chow containing alfalfa was also supplied to provide green (dried) food such as would be normal in the field. Lettuce was given frequently when attempting to breed the woodrats. A variety of fruits, vegetables, and seeds were supplied occasionally. Water was always available in drinking bottles and the rats drank freely, especially in warm weather.

Animals were handled using leather gloves and a hand net or a bag made from 8 mm nylon fishnetting. Tamer individuals could be handled using gloves alone by grasping them firmly around the thorax so that they had difficulty reaching to bite the handler's fingers. Woodrats have fragile tails; attempts to hold them by the tail can remove the sheath almost instantaneously, so that method was discarded early.

Observations were written down in a notebook as the behavior occurred. Other procedures are described in the pertinent sections.

Response to Observer

Recently wild-caught individuals varied considerably in tameness. A few settled down quickly in the laboratory and learned within a few weeks to come to the front of the cage for bits of fruit and seeds. An equally small number never settled down; these literally bounded off every wall of their cage whenever someone entered the room or moved suddenly. Most showed behavior between these extremes: although previously active, they would hide in their nests for long periods after a person entered the room.

Members of a few early litters were handled extensively as nestlings in order to get growth records (reported in field section of this report). These young remained tame until about three or four weeks old. At this age they became strong enough and willing to deliver a painful bite and run away. From this time until several months old they became wary and threatened any hand that approached. After this recalcitrant period they became the tamest animals ever kept in the laboratory, could be handled with near impunity, and became persistent beggars for tidbits, although still nervous and quick to flee.

The tamer young generally were offspring of tame parents. This effect was most dramatic at the opposite extreme: parents which failed to habituate well in the lab almost unfailingly produced nervous offspring which I was unable to tame. Among offspring of tame parents, those not handled as nestlings did not become as tame as their handled siblings.

The refractory period, when captive animals lose their tameness temporarily, falls exactly during the period when the young would be dispersing in the field. The only case I knew of in the field where a bushy-tail became trap-shy occurred at this age. She was recaptured regularly the next season as a breeding adult.

Young from more than one pair of N. cinerea parents demonstrated the refractory effect. N. lepida young, handled similarly, became even tamer, and did not go through such a refractory period.

INDIVIDUAL BEHAVIOR

Posture

Sleeping animals vary their posture depending on temperature. In cooler weather they curl into a ball, either with the feet underneath or lying on one side; the tail is curled around the face. As temperature increases they progress to lying belly down on the substrate with feet extended, and in the warmest weather may sleep on their back, exposing the venter. J. H. Brown (1968) and Brown and Lee (1969) found woodrats resistant to low temperature, but poorly adapted to heat; insulation of the nest and house was important to their thermoregulation in the field.

A common behavior seen in mildly alerted animals at the behavior station was "watching" (plate 4a). The rat would leave the house and lie for long periods on a high ledge. Forefeet were partly extended, with the chin resting on or between them, and hind legs under the animal. Eyes were kept open, and the animal was motionless, but from this position it could move quickly if I moved toward it or if it were otherwise startled. In the field, numerous high ledges are thoroughly urine-marked (plate 5a), indicating that rats climb to them, so possibly this watching behavior occurs in the field, allowing them to keep track of predators. Certainly many of the marked ledges would be difficult for terrestrial predators to get to.

Locomotion

When bushy-tails are moving slowly, normal gait is the diagonal walk common to all quadrupeds (Gray, 1953). When moving quickly they use quadrupedal saltation (cf. Eisenberg, 1963). The extreme of this gait is seen when one is released on a noisy substrate such as dry leaves (rustling leaves or paper create a dramatically frightening sound for woodrats), where they hop in an erratic manner as much as half a meter into the air in escaping.

Bushy-tails are excellent climbers. They had no trouble climbing vertical or even inverted hardware cloth surfaces, from which most could hang for long periods. A few more acrobatic individuals learned to do flips between top, front, and back of office cages, made of hardware cloth. The most remarkable climbing feat was demonstrated by several individuals which learned to climb the two-meter high 90° corners of behavior station cages, where walls were of slightly roughened hardboard. In the field, released animals were seen to climb vertical rock faces as high as five meters, and improbable locations of many urine marks gave further evidence of their climbing ability. The most common method of climbing is to extend both forefeet alternately and then quickly bring up both hind feet together. When climbing is difficult, various modifications of the diagonal gait appear—or, at worst, rapid movements best described as scrambling.

N. lepida and N. cinerea showed similar gaits, but N. fuscipes provided an interesting contrast. The latter species is considered the most arboreal woodrat, and when climbing uses the diagonal gait almost exclusively. When comparing it to the two rock-dwelling species, one is impressed with the smoothness of its motion compared to the jerky movement of the other two. Possibly selection has chosen the diagonal gait in the arboreal rat because it maintains more continuous contact with a less stable substrate.

Bushy-tails apparently rely strongly on kinesthetic memory for escape routes. Animals first released into observation cages were easily confused and recaptured by hand, but could get away quickly after a few days' experience in the cage. If shelter materials or ramps to nest boxes were moved just a few inches, escaping rats previously acclimated to the cages would run off into space or into objects several times before modifying their routes. Learning their routes this way is reasonable since the species frequently nests in the complete darkness of caves (Finley, 1958). The depths of rockslides, particularly with heavy snow cover, doubtless are equally dark. Bushy-tails have particularly long and well-developed vibrissae which probably aid in their dark habitat. Both J. A. Davis, Jr. (1970) and Kinsey (1976) commented on similar kinesthetic orientation by the cave- and rock-dwelling N. floridana magister. Although N. fuscipes rarely lives in rocks, Linsdale and Tevis (1951) observed disorientation in released animals until those individuals had found an obvious woodrat path.

Nest and Houses

Woodrats in behavior-station cages given a choice of nest boxes on the floor or up on the wall invariably chose elevated sites. Although bushy-tails are less known for their houses than other woodrats or packrats (Finley, 1958), when provided with sticks they did a creditable building job. They were able to carry sticks, exceeding a meter long and two to three cm in diameter, up a vertical, two-meter-high wire-mesh wall. At the house the stick was wedged between other sticks, slats of the box, and the wire mesh, so that dismantling a nest was frequently a complex task. Manipulating and carrying of objects were done largely with the teeth, although the forefeet aided in manipulating smaller objects or securing the object in the teeth for carrying.

For the nest cup itself, teeth were used to peel bark off twigs and shred it. Paper and cloth were cut into strips about half a cm wide and several cm long. These pieces were then incorporated into a circular cup just large enough in diameter for the animal to curl up in, and equal to about half the curled height. Domed nests, seen by Finley (1958), were built only rarely. Woodrats maintained in mouse cages with wood shavings usually built a similar nest from the shavings.

Finley found nests built without houses deep in Colorado caves, and I found similar nests in lava tubes at Lava Beds National Monument in California (see Nelson and Smith, 1976). I have also seen such a bare nest of N. lepida in a cave in Nevada, and J. A. Davis, Jr. (1970) described and illustrated similar nests for N. floridana magister. The cue for house building may be excess light at the nest site, since all my laboratory animals built houses when supplied with materials, even though already well sheltered from temperature extremes and precipitation.

The importance of the house in protection from predators is demonstrated by the rats' quick return to it whenever disturbed. This behavior also occurred in the field when I approached animals too closely while radiotracking them. Even when I dismantled houses in the behavior station cages, the animal remained in the nest until the last possible moment. It required effort to induce a rat to leave an intact house.

Food scattered in cages was invariably collected by the residents. Most commonly it was stored within the house, but often separate caches were set up elsewhere in the cages, usually within or against larger objects. Once cached, food was frequently covered with material like sticks, stones, or leaves, resembling a miniature woodrat house. When animals, such as pairs or sibling groups, were caged together, some care had to be taken to prevent a dominant animal from collecting all the food, preventing others from feeding. When this problem arose, I placed food pellets in wire racks that allowed feeding but not removal of pellets. This behavior suggests that theft of food caches by dominant animals may be a factor in spacing of woodrat houses in the field.

Elimination

Woodrats eat a lot of leaves, and consequently have to cycle large quantities of food to be adequately nourished. Much of the great volume of feces produced accumulates in the house, as well as at any site where the animal spends much time. In the field, windrows of fecal pellets may be found in slightly protected parts of their home rocks, and deep layers in the most protected places.

Coprophagy, or reingestion of feces, is well known in rabbits and has been observed in a few rodent species. It has generally been interpreted as aiding digestion of high-cellulose diets in nonruminants by symbiotic microorganisms (Thacker and Brandt, 1955; Moir, 1968; Boley and Kennerly, 1969; Kalugin, 1974). Although Neotoma have diets high in cellulose, coprophagy had been reported only in N. fuscipes (Linsdale and Tevis, 1951). Both N. cinerea and N. lepida were seen to reingest feces regularly during the present study. A characteristic behavior pattern is used in reingesting. The animal brings its anus forward, so it is sitting on its rump. As each pellet is voided, it is tested with the incisors and possibly the tongue. If the pellet is rejected and not eaten, it is thrown as far away as 30 cm by an upward flip of the head. While the animal is resting, this action may be repeated many times.

In smaller cages, urine was normally voided in a corner away from the nest cup. In the largest cages, much of the urine was voided on or near the house, but not in the nest. The paths most used by the rats were also well coated with urine.

POTENTIAL SIGNALS

Apparent sound and odor signals are described here in some detail. Possible postural signals are not described separately, but are included in context in the sections describing interactions, reproduction, and development.

Acoustic (nonvocal)

Two nonvocal sounds are heard frequently when handling and watching bushy-tails. Foot-thumping is a steady rhythmic tapping by one or the other hind foot against the substrate at a rate of 30-90 per minute. Tooth-chattering is accomplished by rapidly moving the lower incisors laterally back and forth against the uppers in bursts of 1-3 sec, or more continuously. These two sounds are heard together (sometimes in alternation) from woodrats in live-traps or backed into a cage corner by a net. Tooth-chattering alone is a normal component of intraspecific aggressive behavior. It may have been derived from an intention movement for biting. Apparently it serves as a threat. Foot-thumping with a hind foot, as suggested in a general context by Ewer (1968), may represent an intention movement for escape. Since foot-thumping may be seen in alerted or cornered animals without tooth-chattering, but rarely occurs in aggressive contexts, her suggestion is apparently supported. The signaling function of both these sounds is suggested further by the fact that animals in separate cages, which can hear but not see others, will often tooth-chatter or foot-thump in response to hearing another do it.

Another sound, tail-rattling, has been heard in many other species of woodrat (e.g., Linsdale and Tevis, 1951, for N. fuscipes). This occurs only rarely in N. cinerea, as when animals are aroused in late stages of aggressive encounters. Development of the bushy tail may have selected against that behavior as a signal, since tail-rattling creates a sound by hitting the tail against other objects.

Acoustic (vocal)

In captivity, vocalizations are infrequent, particularly compared to the two nonvocal sounds just described. Yet a variety of vocalizations occur.

Buzz. Egoscue (1962) described a buzzing sound which I heard often given by the male just prior to copulating. It is best represented by "pss-pss-pss-pss-pss...,' at the rate of about two per second, in bursts of several seconds. Females may also give this sound when sexually aroused (heard from a female that was trying to get into a cage with a male, and which displayed lordosis in spite of little activity by the male—see section on reproductive behavior). In a different context, females emitted this or a similar sound when nestlings were removed for weighing and the female could not reach them through the wire cage front.

Fending. When a female is harassed by a male she may face him, raise one forefoot and her chin, and give a short squeal. Except with highly aggressive males, this causes the male to back off and leave the female. A similar behavior may be seen between immatures when play is too rough for one of the participants. Wallen (1977) described a similar vocalization by N. fuscipes.

Loud squeal. Under extreme conditions, a woodrat may produce a loud prolonged scream that can be heard for a considerable distance. An animal losing a fight may squeal, or this sound may be made by an animal, usually an older subadult male, that has been live-trapped and is being handled for the first time. (They tend to become less agitated with repeated trapping.) The sound is generally accompanied by violent struggling and by attempts, sometimes successful, to bite.

Nestling calls. Juveniles give a piping or chirping call, particularly when not attached to the mother's teats (the "abandoned cry" of Eisenberg, 1959, 1963). This call appears to aid the mother in finding a young which has been dropped or has strayed from the nest. I tested this by placing young at different locations in the cage; the mother often stayed in her nest until the young began calling. When the youngster was placed behind objects, the mother had difficulty finding it unless it was calling,

The young may also give ultrasonic calls, as do many other young rodents (reviews by Noirot, 1972; Sales and Pye, 1974; A. M. Brown, 1976). Often nestlings move their mouths in a manner akin to the piping call, but make no humanly audible sound. Mothers may leave the nest and approach such youngsters, as described above, even when the young are out of sight.

Scent Marking

The morphology, repertoire, and motor patterns associated with scent marking are described here. Additional observations on their context are included in the sections on elimination, adult encounters, and reproductive behavior.

Ventral gland. Aside from the size dimorphism already discussed, the other major feature of sexual dimorphism in N. cinerea is the ventral dermal gland (plate 4b). Males in this species have an extensive area on the ventral surface which secretes a brownish, waxy, musky-smelling material. Howell (1926) and Reynolds (1966) studied this glandular area histologically, and found it to be a field of well-developed holocrine sebaceous glands. When only weakly developed, as in subadults, it is a bare area about 0.5 cm wide by 5 cm long in the ventral midline of the animal. When well developed, in adult males, the bare area in the midline doubles or triples in width, and the brownish stain from the gland extends over most of the ventral surface of the animal in a diamond shaped pattern. In these strongly secreting animals, the glandular skin becomes thickened and develops transverse folds.

If the fur on the back or sides is fluffed back, small dots of secretion (<1 mm) can be seen over most of the body. In spite of this, bushy-tails denied sandbaths do not develop the matted and greasy appearance common to heteromyid rodents (kangaroo rats and pocket mice) under such circumstances (cf. Quay, 1953; Eisenberg, 1963).

As noted by Egoscue (1962) and Reynolds (1966), the gland secretes most heavily during the breeding season. This is reasonable since mammalian sebaceous glands respond to high testosterone levels (Strauss and Ebling, 1970).

Females of _N. c. alticola_ (both laboratory and field animals) show some development of the ventral gland, particularly when nursing young, although it is never so well developed as in males. Nestlings tend to acquire the musky odor from the female's ventral gland. Since females either fail to search for or have difficulty finding lost young which are not vocalizing, the odor would seem unimportant for locating young. Whether it aids in recognition of young was not determined. Females may secrete in response to irritation from the young, in response to hormones of pregnancy and lactation, or both.

Two young males of _N. c. alticola_ showed the ventral stain as early as 38 days, when they weighed 151 and 152 gm.

Egoscue (1962), studying _N. c. acraia_, did not find the stain until 51 days (221 gm) in males. His females never showed any trace of ventral staining. _N. c. acraia_ lives to the south of _N. c. alticola_, is smaller, and less sexually dimorphic, so it is interesting to find differences in this sexually dimorphic feature also.

Scent marking with the ventral gland was first described by Egoscue (1962). It occurs primarily in males, although rarely I have seen females do it. The odor is readily apparent to a human observer who picks up and smells a marked object. Marking is accomplished by the animal's lowering its chest against the object to be marked and walking over the object, rubbing the gland on it. Unless the object is fairly high, the legs are typically sprawled out to the side. The hindquarters may be depressed also, rubbing with the anogenital region, and urine may be deposited. A flat surface, such as the floor of a cage, may be marked, but more commonly a slightly elevated object, such as a stick or the edge of a larger object, is chosen. The animal may mark as much as half a meter at a time. The most common situations for marking are encounters with other males or with odor of other males, and occasionally in precopulatory behavior. Objects marked previously by other males are particularly attractive for marking.

Urine marking. Fresh woodrat urine is milky in appearance and has a mildly musky odor which differs from that of the ventral gland. After it has been exposed to air for a time, especially if it remains damp, it develops a strong skunk-like odor.

Schmidt-Nielson (1964) and Shirley and Schmidt-Nielson (1967) found that N. albigula, as well as other cricetines, can absorb large quantities of calcium from food and excrete it in the urine; the milkiness of the urine comes from calcium compounds precipitated in the bladder. Extensive white deposits seen on rocks in bushy-tail areas are built up by repeated urination in the same spot; rain washes out much of the organic matter, leaving behind the less soluble calcareous material (plate 5).

Such urination spots may be used by many generations of woodrats, building up thick layers which persist for many years after the area has been deserted. If an area is deserted for some time and then reoccupied by new individuals, the settlers almost immediately start leaving streaks of yellow or brownish fresh urine on the old white deposits. This behavior is useful, for it provides a check on occupancy of areas (see field section). Males and females of all ages beyond the nestling stage apparently urine mark in the field. I found fresh urine marks in areas containing solitary individuals of all classes. The amount of fresh marking roughly indicated population levels of the areas.

To mark, the animal simply walks along an object or area, releasing urine as it walks. The hindquarters may be slightly depressed. In males, urine marking can accompany ventral gland marking. The result is a streak of urine about 0.5 cm wide and from 1 to 30 cm long. The streak is usually wavy because the hindquarters move from side to side in walking.

As is common to many animals, urination also occurs in pursued and captured animals; this resulted in a number of unwelcome showers for me while handling woodrats. Kleiman (1971) and Eisenberg and Kleiman (1972) suggested that such urine spraying in some mammals may function as a warning to other conspecifics.

In the laboratory, the most common places for urine marking were along frequently used paths. Cinder blocks left in behavior station cages and rocks in office cages provided sites where bushy-tails duplicated the white deposits seen in the field. Animals in smaller cages simply backed into the corner, lowered the hindquarters, and urinated without walking.

Sandbathing. For practical reasons, woodrats in the laboratory were not supplied regularly with pans of sand, but when these were supplied, vigorous sandbathing often took place, scattering sand widely. The most notable occasion for sandbathing was precopulatory chasing, in which both individuals regularly alternated in the sandbath. Sand left in cages for a day or more became soaked with urine, thus strong smelling, and could no longer be used by the rats for sandbathing.

Two patterns characterize sandbathing. One is similar to ventral gland marking except for the substrate. The animal flattens its belly against the sand, pushing and pulling itself along with its extended feet. The second pattern is a side rub in which the animal turns onto either side and quickly pulls itself forward; the side of the face and the flank are the main body parts rubbed. This side rub may follow a ventral rub, but often occurs alone. Both these sandbathing motions are similar to those of the only distantly related heteromyid rodents, as described by Eisenberg (1963) and which I have also observed.

Anal gland. Howell (1926) described an anal gland in *Neotoma*. His only male *N. cinerea* had slightly larger glands than did his two females, although he was unable to find ducts for the male's glands. In my field and laboratory work, while noting reproductive condition, I observed considerable swelling around the anus of large males, but females showed hardly any such swelling. Gentle squeezing of this region in males produces a slightly viscous clear liquid from the anus. In view of the season his animals were collected (August), and certain features of his anatomical descriptions, I suspect Howell's male was a subadult and its glands insufficiently developed.

The secretion is odorless to me, but an anogenital rub is a common component of ventral gland marking. The gland is clearly more developed in males than in females, so it may have a communicatory function. Mykytowycz (1962, 1965) found an apparently odorless chin gland secretion of rabbits (Oryctolagus) important in their communication. Occasionally I saw both males and females rub the anogenital region against the floor of the cage or onto objects without rubbing the ventral gland, on first being introduced or just prior to copulation. Dixon (1919) erroneously attributed the musky odor of N. cinerea to anal glands.

"Kiss." Bushy-tail mothers and young, when meeting away from the nest cup, frequently touch their mouths together. Such behavior almost invariably precedes retrieving of young by the mother. Less commonly, I saw a similar behavior when a pair of cohabiting adults met. King (1955) suggested that a similar behavior in prairie dogs (Cynomys) functioned as an olfactory recognition cue, and the context here would suggest a similar function.

Other skin glands. Whipple (1904) described glands on the sole of the foot of N. floridana. My N. cinerea left footprints on smooth surfaces, such as plexiglass doors of office cages, but I could not tell whether these came from glands or from tracked urine. It is unclear whether any communicatory function is involved; in such a good climber, keeping feet soft and moist might be important.

Preputial glands function in olfactory communication of other mammals (Bronson and Caroom, 1971; Brown and Williams, 1972, Mugford and Nowell, 1971). Prior anatomical studies have found no preputial glands in most Neotoma species (N. floridana: Arata, 1964; N. fuscipes, N. albigula, N. cinerea, N. micropus, N. floridana: Howell, 1926). In spite of these earlier findings, I made preliminary dissections of adult male N. cinerea and found what appear to be paired preputial glands containing a viscous material. As with the anal glands, possibly Howell's use of a subadult specimen resulted in his missing poorly developed glands. Howell did find reduced preputial glands in N. lepida (identified by him as N. intermedia, a name now applying to only a subspecies of N. lepida). The preputials of N. cinerea could add odor both to urine marks and to ventral gland marks, for both behaviors often include an anogenital rub.

INTERACTIONS BETWEEN WOODRATS

Male-Male Encounters

Field and museum evidence indicated that males exclude each other from breeding areas. Ralls (1971) had pointed out that mammals frequently mark in aggressive contexts, and male N. cinerea possess strikingly sexually dimorphic scent glands, so one might predict that encounters between adult males would be aggressive and involve scent marking.

I arranged encounters in a number of ways, either introducing two males to a neutral cage or introducing one male to a cage containing another which had been resident for a considerable time. Both office and behavior-station cages were used, but the office cages were usually too small, and little behavior could be seen besides scuffling and running. In the larger cages a fairly consistent sequence of behaviors appeared, which is described below. Serious fighting was the inevitable outcome, and I usually felt constrained to stop the fighting, to preclude damage to participants, before any clear dominance had been established.

Whether both rats are new to the encounter cage or one is an intruder, behavior is similar. In either case, a rat new to the cage sniffs the air and substrate continuously while cautiously exploring the cage. On detecting the other rat by scent, sight, or sound, intense tooth-chattering is begun. The ears, which had been brought slightly forward to an alerted position while exploring, now may be directed stiffly ahead. Unless a resident is sleeping when the action starts, tooth-chattering is quickly answered by the second animal. Before engaging in actual fighting, both individuals normally scent mark on various objects with the ventral gland and with urine. Often ventral gland marking is preceded by the animal sitting upright and vigorously using its teeth to groom or chew at its own ventral gland. This behavior could perform several functions: bringing more sebaceous secretions to the surface, or wetting the ventral gland and adding salivary odors to it. Ewer (1968) and Eisenberg and Kleiman (1972) have suggested that an animal might gain confidence or self-assurance from its own odor, which could be a factor here.

Following these preliminaries, which may take place at distances of 2-3 m, either woodrat may simply rush at the other, leap at it, and begin fighting. More commonly, however, initial approaches are slower, the animals cautiously moving together tooth-chattering, sniffing, and taking time to scent mark objects along the way. When close, both animals stand up on the hind legs, face to face, and place forelegs against the other animal. (This appears more a stereotyped posture for remaining mutually upright than the "boxing" frequently described for other rodents, for there is little actual pushing or sparring by the forelimbs.) Each animal stretches its head as high as it can, nose in the air, both tooth-chattering intensively. From this position, each tests the other by cautiously lowering its head, seeking an opening.

So long as one animal has its head up, the other appears inhibited from attacking. As soon as one head is lowered, the other animal may try to bite at ears, neck, or face. A common pattern is for one to start lowering its head, see a counterattack coming, and immediately jerk its head upright again, forestalling the counterattack. The mutual upright ends when one either makes a serious attempt at biting or leaps at the other, attempting to rake it with the claws of the hind feet. The two then back off, go through the ventral marking routine again, and repeat the performance.

After several bouts, either may crouch briefly, in preparation for rushing the other, and vibrate its tail laterally. At this stage, the mutual upright posture may be abandoned, and each alternates scent marking with rushes followed by a leap at the other. Frequently hind feet are used to scratch; they may produce cuts several cm long on the side or belly of the other animal. (In such an adept climber the claws are strong and sharp, requiring me to use gloves in handling even the tamest rats.)

Unless animals were grossly mismatched, one or both animals frequently received serious wounds before either reached the point of fleeing. This may have resulted from the small space available even in the largest cages, for I only occasionally saw much wounding in the field population, with the exception of a small number of cut ears or bitten rumps. In the laboratory, the most common wounds were bites on the ears and face from the initial upright phase. Bites to limbs, whether on the feet or closer to the body, proved the most troublesome, and caused some animals to limp for a few days to several weeks afterward. Occasionally, aggression developed between mated pairs or groups of siblings caged together. Then a subordinate animal, in attempting to flee, would be bitten repeatedly on the rump. The woodrats were remarkably resistant to infection; only occasionally did wounds become infected, and these usually responded to a tetracycline preparation dissolved in the drinking water for a few days.

Prior residency in the cage seemed primarily to affect the ability to get around in the cage, as described above in the locomotion section. An intruder was less able to escape quickly or to find hiding places. Where both animals were otherwise evenly matched in size this might have provided a slight edge for the resident. Otherwise, the probable outcome, if fights had been allowed to continue, seemed to depend on how closely matched the combatants were in size.

The ventral gland marking in the fight can be seen as a component of agonistic display (cf. Ralls, 1971). Males presented with odor from the ventral gland of another male, in the absence of the other animal, responded by tooth-chattering and often by grooming their own ventral gland and by marking. This behavior could be elicited by odors from sticks or paper rubbed on the chest of another male, organic solvent extracts of material combed from the chest, and odors from a male in an adjacent cage which was quiet and not visible. Eisenberg and Kleiman (1972) indicated that such aggressive responses to the scent of an absent conspecific are not commonly observed in mammals. Unfortunately, the response was not sufficiently consistent to carry out more detailed experimental studies as originally intended. "Nervous" animals generally failed to respond, and even the tamer males, which usually responded well, would "freeze up" when faced with a new situation. The response is real, but is easily masked by fear of other factors in the situation.

Female-Female Encounters

For comparison with male-male behavior, a small number of female-female encounters were staged. In short-term introductions, fighting may occur, but without the intensity of male encounters. Placed in a neutral cage, females spend more time investigating the cage than each other, and there is a considerable period before they even approach each other. No ventral marking occurs, but there may be urine marking and dragging of the perineal region.

First approaches are tentative, sniffing nose-to-nose, ears back, including some tooth-chattering, and usually followed by simply moving away. Eventually, brief mutual uprights occur, but these include more boxing and pushing than seen between males, and are broken off more quickly. Bouts of fighting are brief. Even after serious attempts to bite occur, the rats spend much of their time away from each other. The impression, in contrast to males, is that each female is more ready to avoid the other than to fight, but that neither would be tolerant of the other in the long run, at least in the space provided.

REPRODUCTION

Bushy-tails were bred in the laboratory for three reasons. (1) Since olfaction plays a role in mating behavior of many mammals, I hoped to assess its role in this species. (2) Because of problems in trying to observe fearful wild-caught animals, laboratory-raised young offered the possibility of more tractable subjects. (3) Finally, observation of growing young woodrats, particularly interactions among themselves and with parents, might provide useful clues to events in the wild.

Breeding Procedures

Most breeding was carried out in the larger office cages where surveillance could be maximal, particularly when pairs were first introduced. Many attempted pairings failed due to fighting, and the individuals had to be separated to prevent death, usually of the female.

Woodrats were generally not put together for breeding unless both partners appeared physiologically ready. For females, this meant the vagina had to be open, and tissue surrounding it vascularized to a pink or even purplish color. In some cases, females even prior to breeding had bare skin surrounding the nipples and were seen grooming this region extensively. Male Neotoma when fully scrotal are not always so obvious as are common laboratory Rattus or Mus, and may retract the testes when picked up for examination, so manipulation with the fingers is generally necessary.

Once judged ready, the two animals were released at nearly the same time into one of the larger office cages in which neither had been living. To break up the space a bit, create some visual barriers, and eventually to provide nesting material, I often placed large quantities of loosely crumpled paper into the cage beforehand. Since intended pairs frequently fought, it provided some temporary shelter. Introductions were generally made in the morning so that I could observe for most of the day and evening and be available to rescue an extreme underdog. If pairs fought excessively, they were separated; some of these were later successfully mated when reintroduced either to each other or to other mates. Behavior seen is described in a separate section below.

So far as possible, pairs that "got along" were left together for the breeding season, which extended in the laboratory from February through July. "Getting along" ranged from pairs with an uneasy truce that could erupt into fighting at any time to pairs in which I almost never saw an indication of fighting. A number of pairs, once they proved capable of living together peacefully, were transferred together to behavior station cages. In some cases, original introductions were made at the behavior station.

Egoscue's (1962) useful suggestions for breeding *N. cinerea* were followed, with a few exceptions. I did not provide nest boxes in office cages because I found that when I did, I rarely saw the animals. The woodrats would retreat to the boxes and failed to become habituated to my presence at all. They were less likely to leave nest boxes when I was present than they were to leave open nests. Fighting between mated pairs tended to increase when nest boxes were provided, a problem I later found that Birney (1973) had encountered with *N. floridana* and *N. micropus*. Finally, open nests allowed me to see more of mothers with early young.

Young were left with parents for periods up to two or three months after birth because I wanted to observe any changes in their interactions with parents and with each other. In a few cases, when crowding was a problem, I removed siblings together to a separate office cage until I was later forced to separate the siblings.

Breeding Records

In table 9 are summarized the successful pairings and births in the laboratory. A number of other unsuccessful, or at least unproductive, pairings were attempted; in these, pairs were together for periods ranging from a few minutes to nearly six months (typically only a few days). Although a few pairs started living together in January, no births occurred prior to mid-March. I left pairs together into August, but there were no births after late July. The breeding season both started and ended about two weeks earlier than in the field population at Sagehen Creek (cf. table 15). As Egoscue (1962) noted, pairs tend to start fighting in July and August. My observations of this were incomplete because I spent most of August and September each summer in the field and had to leave care of my animals to others.

Egoscue described a gestation period of 27 to 32 days. Chapman (1951) reported for N. floridana an estrous cycle of 4-6 days, which was later confirmed by Birney (1973). A frequency distribution (fig. 3) of birth intervals from the present laboratory colony agrees generally with these sources. It further suggests that the initial postpartum estrus

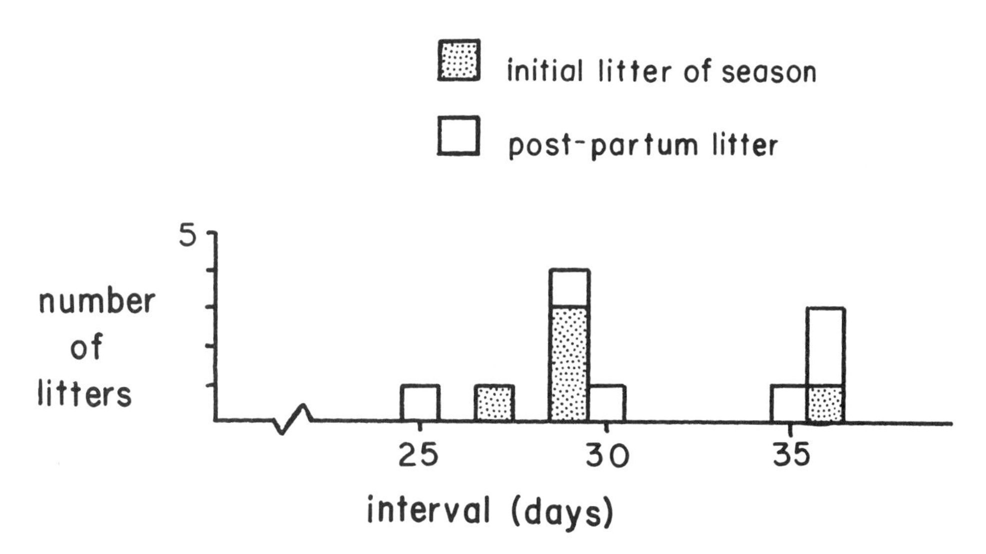

Figure 3. Birth intervals of Neotoma cinerea in the laboratory: time from introduction of pair until first litter, or between successive litters, when interval was less than 40 days.

TABLE 9
Births of Neotoma cinerea in Laboratory

Pairs F x M	Date	Number of Young Total	M	F	?	Interval (days)	Notes
3 x 1	17 Feb '71						pair together
	24 Mar	4	2	2		36	
	29 Apr	3	2		1	36	1 born dead
3 x 1	26 Mar '72						pair together
	24 Apr	3		3		29	
	20 June	2	1	1		(50)	
	24 July	2		1	1	35	parents killed young, 5 Sept.
3 x 1	1973	*					*
6 x ?	23 June '71						caught pregnant in field
	6 July	4	3	1		(23)	
6 x 2	21 Feb '72						pair together
	22 Mar	2		2		29	
	27 Apr	1		1		36	
	26 May	1		1		29	
	21 June	1	1			25	
6 x 2	1973	*					*
19 x ?	11 May '72						caught pregnant in field
	3 June	5	2	3		(23)	all weaned
19 x 14	21 Jan '73						pair together
	14 Mar	2		1	1	(53)	1 born dead
22 x 11	7 May '73						pair together
	3 June	2	1		1	27	both dead at 4-5 days
	4 July	3	1	2		30	
24 x 7	29 Mar '73						pair together
	27 Apr	2		1	1	29	dead at 2 and 8 days
32 x 28	19 Mar '73						pair together
	1 June	2	1		1	(72)	1 dead at birth
	Totals	39	14	19	6		16 litters

*In 1973, 3 x 1 were together from 17 Jan-29 Mar and 5 July-25 July, with no young produced; 6 x 2 were together 22 Jan-5 Aug, with no young produced.

may be missed, or not taken advantage of, but that the female immediately starts a new series of 5-7 day estrous cycles. Egoscue pointed out, and present data corroborate, that lactation does not lengthen gestation in this species. Based on initial gestation periods for several females (table 9), many pairs must have mated within a few days of being introduced.

I have not given here the easily calculated figures for average litter size and average number of litters per season per female because I feel they would be misleading and unrepresentative. Litter size in the laboratory declined with time, both for females mated in more than one year (#3, 6, and 19) and for young females first breeding in 1973.

Mating Behavior

Since observation of copulatory behavior was only one of several aims in pairing bushy-tails, no hormone treatments were used to induce estrus (cf. Dewsbury, 1974a, 1974b). As a result, occurrence of copulation was unpredictable, and observations of mating largely depended on spontaneous cycles of pairs in office cages. Fortunately, in spite of the animals' shyness in other respects, they would occasionally copulate in my presence if I sat quietly. I saw at least one or two copulatory sequences for almost all pairs listed in table 9, many more than that for a few, and some sequences for pairs that did not produce offspring. Many of these observations were only of partial sequences, however, for I did not always see the beginning of the activity, my attention often being attracted by hearing the "buzz" vocalization associated with copulation. Sequences also were interrupted often by noises from another room or by inadvertent movement on my part. The description that follows is thus a composite representing the typical pattern and indicating some of the variations.

Initial Interactions

When a pair is first introduced, both start by exploring the cage. Both sexes mark with anogenital rubs in intersexual encounters; males occasionally mark with the ventral gland, but not with the intensity and frequency seen in male-male encounters. Initial approaches are often cautious, with the ears cocked backward, the body low and elongated. Either or both may start tooth-chattering, and chases are common. Most often the male does the chasing, although the reverse may occur, particularly at the beginning of a pairing. Unlike encounters between males, initial fighting is rarely so serious that combatants' lives are in immediate danger. The female (or, less commonly, the male) escapes by running up the cage side and lying on one of the 4 cm wide horizontal crossbars, where it is ignored unless it attempts to return to the cage floor. In unsuccessful pairings, the male keeps the female off the cage floor almost completely and thus away from food and water, requiring her removal after about two days to prevent starvation and/or dehydration. At this stage, severe wounding is uncommon, but animals can die when 25 percent of body weight is lost. These problems are obviously an artifact of the small space in captivity. In the wild the female could live elsewhere and find other resources.

In successful pairings, where animals at least live together peacefully, the female either stops or delays the male's aggression by facing him and giving the "fending" vocalization, refusing to run when chased, or even initiating some chases herself. If a sandbath is present the pair may alternate sandbathing as well as performing anogenital rubs on its edge or on the cage floor. In these successful pairings neither individual succeeds in completely dominating the cage floor, nest materials, or food and water. Separate nests and food caches are commonly located at oposite ends of the cage. To some extent, shifts in dominance can be tracked by observing the distribution of nest materials and food stores, although the latter may be kept in a shared cache (cf. Egoscue, 1962).

Copulation

Although the records of gestation intervals (table 9; fig. 3) demonstrate that copulation must take place within a few days following introduction for most pairs, I never actually saw it occur in that period. I remained by the cage continuously for at least 1-2 hours after introductions, and spent much time there for several days and evenings following. Most copulations observed were thus in the postpartum estrous period—in one case less than 12 hours following birth—or at variable intervals during a nonproductive pair's period together.

As noted by Egoscue (1962), copulations often occur in the afternoon and evening. In the lab, animals tended to become active in the afternoon, although the field populations were almost never active or trappable then. Possibly there is some local activity in or near the houses, under the rock slides, that I was unable to detect.

A typical copulatory sequence begins with the male approaching the female and making the "buzz" vocalization. If the female is in her nest but receptive when approached, she will leave her nest and house.

Chasing is a common prelude to copulation. In the larger behavior-station cages precopulatory chases were frequently prolonged, and actual copulation took place on the cage floor well away from the houses. If sand is present, the animals may alternate sandbathing during the chase. When the female slows down, the male nuzzles her hindquarters and genital region, continuing the vocalization. Actual grooming of one adult N. cinerea by another is rare in this or any other circumstances. She then may stand still and adopt a lordosis posture with the hind-quarters elevated, and the male mounts.

The copulatory position is similar to that figured by Dewsbury (1974a) for N. albigula, with the male resting his weight only lightly on the female, although in N. cinerea the male appears to hold the female more tightly with his forelegs than Dewsbury suggested for N. albigula. After mounting, the male thrusts 5-10 times in 1-2 seconds and simultaneously may rub his chin rapidly back and forth across the middle of the female's back. Chin rubbing was seen consistently with certain pairs, never with others.

Following the male's thrusting, the female runs forward, usually leaving the male standing alone in the copulatory posture, although he may take a few short steps in attempting to maintain contact. After separating, both generally groom the genital region. No evidence was seen of a copulatory lock as described by Dewsbury for N. albigula (1974a) and N. floridana (1974b). The sequence of mounting, thrusting, and separating may be repeated 4-5 times in less than 1-2 minutes, after which both may feed, return to their nests, and ignore each other.

One unusual sequence, although it did not result in copulation at the time, is described in detail because it suggests a number of other features of sexual behavior. Male 28 and female 32 were born in the lab in June 1972, to different parents. They were put together in an office cage on January 19, 1973, but were separated on the 22nd because of excessive dominance by the male. Another attempt on the 24th lasted for only 45 minutes because of fighting. The next day they were placed in adjacent office cages several cm apart. (The cage floors differed in height by nearly a meter.) In the next weeks the two interacted frequently, orienting toward each other, sniffing, and occasionally tooth-chattering. Around midnight of March 18 the female was persistently orienting toward the male and giving the buzz vocalization usually heard from males soliciting copulation. In response, the male was shaking his head and making the stiff-legged sideways hop usually associated with play in young woodrats. Because of their obvious interest in each other, I decided to try pairing them again.

While I was catching the male to transfer him, the female sat up and groomed her chest in a manner similar to that described for males preparing to ventral mark and fight. For the first five minutes after transfer the male explored the female's cage (no marking seen) and the female stayed out of his way by climbing up on the side of the cage.

Eventually she came down, approached the male, and repeatedly solicited copulation for the next 5-10 minutes. Each time, she typically approached the male from the front, sniffing nose to nose for 1/2-2 sec, the male sitting partially upright. She then walked forward turning her head behind and under the front leg of the male as though attempting to reach and groom his ventral gland with her teeth. Next she turned around and assumed the lordosis posture directly in front of the male with her chest depressed to the floor, her genital region elevated, and tail to one side and quivering, holding that position for 1-2 sec or more. Usually the male investigated her genitalia while she held that posture, but he did not attempt to mount. The female dropped her rump, ran a short distance, turned around, and repeated the performance. No vocalizations were heard from either animal.

Finally, the male began to mark with his ventral gland and anogenital region in numerous places around the cage. The female several times went through the motions of sandbathing, although no sand was present, and proceeded to anogenitally mark where the male had marked. By 12:30 a.m. (15 minutes after the introduction) the male was mostly avoiding the female, but had threatened her only once.

At 12:35 I put in a pan containing fine sand, wherein the male almost immediately started dustbathing as well as ventral-marking on the edge. The female appeared hesitant to approach the strange object, but after 20 minutes also started dustbathing, following which the male groomed her head and neck, and she again solicited the male several times. This time he responded each time by licking her genitalia for 5-10 sec, but she always moved away, and no mounting attempts were seen.

By 1:05 a.m. the rats appeared to lose interest in each other and were exploring the cage and feeding, so I left. Subsequently, on the afternoons of March 26 and 30 (intervals of 7 and 4 days), I again saw the female solicit several times in the same way. On these occasions the intensity of the lordotic posture seemed less, and again no mounts were attempted. On May 5, I moved the pair to the behavior station, and the female gave birth to two young 25 days later, on June 1.

I did not see the female of any other pair solicit the male in this fashion, so it is unclear whether the pattern is aberrant, or simply represents a normal pattern brought out by unusual circumstances. Pearson (1952) and W. J. Hamilton, Jr. (1953) observed females of N. floridana which solicited copulation from slowly responding males, but neither case involved such extensive scent marking.

Dewsbury (1974a, 1974b) described copulation of N. albigula and N. floridana, which he brought into behavioral estrus by hormone treatments. He did not discuss scent marking, but his studies were concentrated on other measures. Males of both these species, although having a less developed ventral gland, have been observed to ventral-mark in this or other situations (R. J. Howe, personal communication; Knoch, 1968; Kinsey, 1976).

Nearly every potential avenue of olfactory communication was displayed in the present sequence in a short period of time. Under more normal circumstances, as in the wild, the pair would have lived near each other for several months prior to the breeding season, thus a much longer period would have been available for exchange of olfactory cues. Since I was unable to observe the initial copulatory activity in any other pair, the possibility exists that similar behavior occurred unobserved in those other pairs.

The intensity of scent marking seen here may have been the result of a female in a high state of behavioral estrus confronted with an inexperienced and unresponsive male. Normally, the scent marking might have taken place over a longer period with less intensity, but of no less importance. The "telescoping" here of long-term behavior into a short period may serve to emphasize the potential importance of olfactory cues in mating, which might not have been obvious otherwise.

Comparison to other rodents

Dewsbury (1972, 1975) has observed copulatory behavior in a great many rodent species, particularly the neotomine-peromyscine group to which woodrats belong, and arrived at a number of generalizations based on those data (summarized in Dewsbury, 1975). My observations were conducted under different conditions, and I was unable to observe all of the factors which he defined, but certain aspects can be compared. One of the first features he discussed was "intromission latency," the length of time between introduction of a pair and the first intromission. He suggested that long intromission latencies might be correlated with pair-bonding species (e.g., *Onychomys* (grasshopper mouse) and *Peromyscus polionotus* (old-field mouse), and short latencies with nonbonding species.

The two woodrats Dewsbury studied in detail, *N. albigula* and *N. floridana*, both showed short intromission latencies. Unfortunately, little is known of their social organizations. If the lack of immediate copulation in the majority of my *N. cinerea* (not hormone-primed as were Dewsbury's subjects) can be considered as a long latency period, then my observations in this area are consistent with his predictions. I would expect something approximating a pair bond to develop within the small harems which are maintained for a significant part of a bushy-tail's lifetime. On the other hand, it might not be realistic to compare intromission latencies between these species, for Dewsbury used different amounts of hormone for different rodent species. There may be interspecific differences in dose response. Such differences could be most evident in measures such as intromission latency which depend on both receptivity and attractiveness of the female, although I am unaware of published data to resolve this problem.

Dewsbury found a good correlation between penile morphology (table 10) and the presence or absence of a copulatory lock. Rodents with a short thick penis (glans diameter greater than 29 percent of length) never locked. On the basis of penis dimensions alone he had been able to predict the occurrence of locking behavior in several species. Thus, within Neotoma, he found that N. albigula and N. floridana, with thick penes (table 10), are locking species (Dewsbury, 1974a, 1974b), while N. lepida, with the thinnest penis in the genus is not (Dewsbury, 1975; Estep and Dewsbury, 1976). My observations of N. cinerea reveal that it does not lock, confirming Dewsbury's predictions, since its ratio is only 23.

Dewsbury also suggested that presence of a lock might be correlated with species that could copulate in a safe retreat, such as a woodrat house. Both N. albigula and N. floridana build large houses (Finley, 1958) and are locking species. I would point out further that the locking (thick penis) woodrats are the "best" house builders. N. fuscipes, with the stoutest penis (copulatory behavior uncertain, although Wood, 1935, did not describe a lock for the copulatory sequence she observed), is noted for building the largest houses (Linsdale and Tevis, 1951), and N. lepida, a nonlocking species, builds only small houses (Finley, 1958; Cameron, 1971). The lack of a lock would indicate a danger from predation while copulating. In this regard it is relevant that N. lepida is normally shy about copulating in front of observers, for neither Egoscue (1957 and personal communication) nor I ever observed their copulation in spite of our finding them otherwise much more amenable to laboratory life than N. cinerea. N. cinerea is also not much of a house builder, but by living among large rocks has better cover than does N. lepida, and may act in the laboratory as if it were in a cave or among its home rocks.

TABLE 10
Ratio of Glans Diameter to Length in Neotoma
(after Hooper, 1960)

Species	Ratio
N. albigula	47%
N. floridana	53
N. fuscipes	79
N. lepida	11
N. cinerea	23

Dewsbury also noted that locking species generally show little chasing, which would agree with the concept of copulating in the enclosed space of a house. Where there was space available, considerable running occurred among my N. cinerea, and the animals always left their houses and nests before copulation. These observations suggest that in the wild much of the copulatory activity takes place away from the houses.

Parturition

I witnessed the birth process on three occasions, all near the middle of the day. In two of these there were two young born, one of which was dead at birth, and the process took about one hour. In the other case (the first litter in the lab), there were four young born. At 10:45 a.m. the first contractions were seen, and the female took up a birth position with her back arched and her body upright and raised off the floor by stretching with all four legs. The young were born at 10:55, 11:20, 11:50, and 12:45, each followed in about 10 minutes by its placenta, which the mother ate. By the time the fourth baby had come, the first two were already nursing. The mother groomed the babies and her own genital region extensively. Her mate investigated briefly, but quickly retreated to his corner again without any obvious threat from the female. Midday births may also occur in the wild, for on one occasion I released a pregnant female at 9 a.m. and recaptured her at 1 p.m. no longer pregnant, although the stress of trapping could have altered normal timing. Horvath (1966) also observed a daytime birth in the wild.

DEVELOPMENT OF YOUNG

Numerous workers have described early growth and development in several species of Neotoma (Poole, 1936; Pearson, 1952; Rainey, 1956; Egoscue, 1957, 1962). Two particularly good descriptions are available: W. J. Hamilton, Jr. (1953) detailed morphological changes in N. floridana and included excellent photographs; Richardson (1943) made careful obervations of behavioral development in N. albigula young.

Except for obvious differences due to size and adult pelage color, there appears to be little variation in the first month of development within the genus. Observations from the present study and that of Egoscue (1962) prove N. cinerea to fit the pattern; the previously mentioned papers should be consulted for more complete descriptions of early behavioral and morphological development than are presented here. Molt sequences for N. cinerea have been well described by Finley (1958) and Egoscue (1962). Growth data were recorded in the present study and are discussed in combination with field data in that section of the report.

Woodrats are born nearly naked and with the eyes closed (plate 2a). They spend much of their first three weeks attached to the mother's nipples. The incisors at birth are splayed out laterally, providing a firm grip on the nipple, and an escaping female will drag her attached young considerable distances (plate 4c), as observed by many workers. This adaptation of newborn tooth morphology has evolved independently in several rodent groups (Lawrence, 1941). Although the incisors have grown out and worn down to their adult condition by three weeks of age (W. J. Hamilton, Jr., 1953; and present study), they can still maintain a firm attachment. I saw bushy-tail mothers run up the vertical cage side with four young of this age attached, and whose combined weight exceeded hers. In the nest, the mother detaches the young by turning herself several times in a circle, pushing at the young with her forefeet and mouth. Removing young for weighing was a problem because forceful pulling could lacerate the nipples. Following Egoscue's (1957) suggestion, I found that briefly holding the youngs' nostrils closed caused them to release their grip.

Females retrieve young with their teeth by grasping them around the belly from the ventral side. This retrieval position, as well as resumption of the nipple attachment, is facilitated by the young, who respond to the mother from the first day by rolling on their back and pulling themselves backward to the nipples. As earlier demonstrated by Richardson (1943) for N. albigula, this is a response to tactile stimuli, for it can be elicited by touching the young with a hand or other object.

Between days 15 and 20, considerable changes take place. Juvenal pelage has grown out, eyes open at day 15 or 16, and the young start exploring the cage and trying solid food. Escape and defense reactions are first seen on days 19-20. They can be successfully weaned by day 25, although they generally continue some nursing for another week or two if left with the mother. Arrival of another litter around day 30 does not always end nursing by the first litter, but will at least reduce it. Until the young are eating solid food and moving around the cage, the mother stimulates their elimination by licking the anal region, consuming their feces and urine.

The male has little to do with the young up to this point. Except for a brief period during estrus, the male avoids the nursing female. I never saw a male do more with a nestling than briefly sniff it.

From three weeks to two months of age, much of a woodrat's time is occupied with play. Practically everything in the cage, movable or immovable, is investigated and reinvestigated, usually with the teeth. Movable objects are stacked and scattered, first in one part of the cage then another. Much running, chasing, and play-fighting occurs. Typically, to start mutual play, one young approaches another shaking its head and hopping in a characteristic stiff-legged fashion, which is responded to by chasing or play fighting. Wilson and Kleiman (1974) identified such patterns ("locomotor-rotational movements") as common to many, if not all mammals; the head shake may be related to responses to olfactory stimuli, and the jumping to predator-escape behavior. When play-fighting first appears at three weeks, it involves gentle batting and mouthing of the opponent. At six weeks to two months of age, play turns to more serious fighting, so that I usually had to separate siblings in office cages from each other at about two months of age.

Young also play with both parents, or attempt to do so. Playing young climb and jump on the parents and pull on their tails, ears, and legs. Parents respond primarily by gently batting at or mouthing the young. Often the male will simply close his eyes, lay his ears back, and lower his head, a posture commonly adopted by submissive rodents.

This tolerance by adults seems to last at least as long as, if not longer, than the young's tolerance for each other. Groups of siblings separated from each other at two months (one group of three was kept together for three months) were seen attacking each other regularly and almost invariably had wounds ranging from torn ears to chewed rumps. One male offspring was left with his parents with no problems until he was removed for other reasons at 11 weeks of age. A 10-week old female was seen to steal food pellets repeatedly out of her mother's mouth; she responded only by turning away to try to keep the pellets and then by collecting a new pellet to replace each lost one. This trait, as is true of much woodrat behavior, shows considerable individual variation, for I saw an adult male threaten his male young at seven weeks of age. They were separated at nine weeks, with no obvious wounding of the young having occurred.

Late summer, the end of the breeding season, is a period when some males not only may turn on their mates, but also on their young. The only case where a parent killed its young occurred September 4-5 when a male at the behavior station killed both of his six-week old young as well as severely wounding his mate.

The period of young playing together has its counterpart in the wild. Bushy-tails of this age occasionally were live-trapped together, indicating that they had entered the trap simultaneously. This period, particularly for young males, but also for some of the females, was when dispersal began. Certainly the increasing strife, both with siblings and with parents, would contribute to the young's leaving the home area.

FIELD STUDY

To study a species' normal social organization properly we ideally should observe it in its natural habitat with as little disturbance as possible. As suggested in the introduction to this report, nocturnal rodents in general have proved intractable to such an approach. The survival strategy of a small nocturnal rodent is geared to being inconspicuous or "cryptic." Nocturnality is only one aspect of their cryptic behavior. Most live and forage solitarily, largely restricting their movements and nest sites to sheltered areas.

Kinsey (1971, 1972, 1976) demonstrated that changing the density of captive populations of two different species of woodrats could make dramatic changes in social organization of those populations. Thus, laboratory observations on woodrat species need to be interpreted relative to normal organization in the wild.

In 1976, Kinsey stated in his paper on laboratory social behavior of *N. floridana magister* that "To date ... no successful observations of social behavior under natural conditions have been reported for the Allegheny woodrat or any other members of the genus." For most species of woodrat (and other nocturnal rodents) I suspect we will not be able to make detailed field observations of their social *behavior*, because in most cases dense cover of their habitat makes them too hard to see, probably even with night viewing devices. Cave-dwelling populations of *N. floridana magister* and *N. cinerea* may prove to be exceptions, for J. A. Davis, Jr. (1970) and Nelson and Smith (1976) described these as relatively fearless and amenable to observation. On the other hand, in spite of numerous field studies conducted on various species of woodrats, few workers have addressed the question of social *organization*, except to note regularly that houses are almost never occupied by more than one adult.

The most thorough study of a single species, that by Linsdale and Tevis (1951), of the dusky-footed woodrat (N. fuscipes), approached the question by documenting changes in house occupancy over a long period of time. Their discussion of social organization (pp. 602-620), however, is difficult to interpret. It alternates between generality and detail, with little middle ground. The social organization cannot be reconstructed clearly from the data given, which is unfortunate considering the volume of data they collected. Following each woodrat over a long time seems to have caused these workers to see them as individuals, which was both their strength and weakness. Linsdale and Tevis pointed out, as has Leyhausen (1965) more recently in another context, that solitary living for a mammal still implies a considerable degree of social organization and interaction among neighbors. They emphasized the role of individual variation and temporal changes in controlling social organization in a particular time and place, but general principles are lost in the welter of examples given. Their work preceded by some time our current interest in theories behind social organization, but they were also reluctant to speculate. They lacked information on short-term movements (as within a single night), which might require radio equipment unavailable then, as well as more detailed information on behavior which would be available only from laboratory studies.

Since not much behavior can be seen in the field, and normal organization cannot be inferred safely from laboratory studies, the two approaches must be taken together. Observations from the laboratory can aid in interpreting events in the field, and vice versa. Wallen's (1977) study of N. fuscipes exemplifies that approach. By studying the same individuals in both lab and field he has provided insights into that species' organization which neither Linsdale and Tevis' (1951) field study nor Kinsey's (1971) lab study of that species provided. Cranford's (1970, 1977) radiotracking study of N. fuscipes dealt primarily with home range, although he hypothesized a temporary pair bond during the breeding season based on changes in home range overlap at that time.

The field portion of the present report on Neotoma cinerea is important to an understanding of its social organization, for little information on that subject is available in other field studies of the species. Finley's (1958) detailed study of distribution, nest sites, food habits, and other aspects of bushy-tail natural history did not address itself to questions of social organization. He included the only report of a mark-recapture study for the species, but it was not discussed in sufficient detail to be useful for present purposes. The only published reference to social organization in N. cinerea is Dixon's (1919) statement that a single rock slide of "two to five acres" contains only "one family of five or six individuals." Banfield (1974) repeated that statement in his account of Canadian bushy-tails; comparison suggests that he was relying heavily on Dixon's study of California bushy-tails for much of his account, even though Dixon was not cited in his references. Local and regional faunal treatments also contain details on N. cinerea natural history, most of which draw on each author's own field experience with the species (Bailey, 1918, 1930, 1936; Grinnell and Storer, 1924; Grinnell, Dixon, and Linsdale, 1930; W. B. Davis, 1939; Dalquest, 1948; Sumner and Dixon, 1953).

STUDY AREA

The Sagehen Creek Field Station (fig. 4), maintained by the University of California, provided an excellent base for field study of N. cinerea. It is located on the east slope of the northern Sierra Nevada, at an elevation of 6400 feet (1950 m), eight miles (12.9 km) NNW of Truckee, Nevada County, California (see fig. 2). Climax vegetation in Sagehen Creek basin is of the Yellow Pine/Jeffrey Pine type (Storer and Usinger, 1963), dominated here by Jeffrey pine (Pinus jeffreyi), white fir (Abies concolor), and lodgepole pine (Pinus murrayana). The eastern (lower) portion of the basin was burned in 1960, and supports an early successional brush field dominated by tobacco brush (Ceanothus velutinus) and greenleaf manzanita (Arctostaphylos patula). More detailed descriptions of plant succession in the Sagehen area may be found in the papers of Bock and Bock (1969) and Beaver (1972).

Because N. cinerea depends on rocks for shelter, its occurrence is closely tied to the geology of the area. Most rock in the basin is

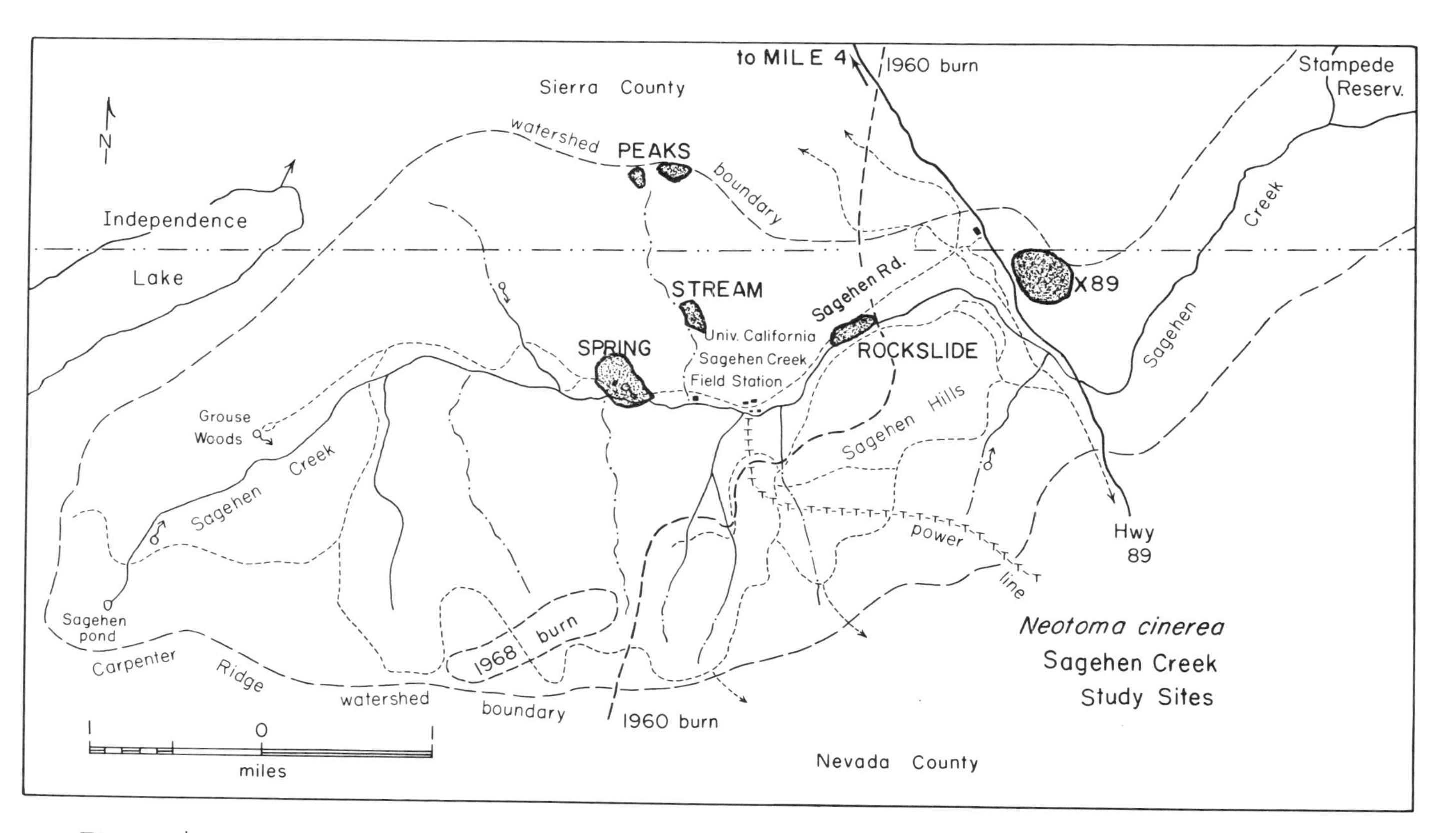

Figure 4. Map of Sagehen Creek basin, showing main study sites for *Neotoma cinerea* (stippled areas). Short-dashed lines indicate dirt or gravel roads. Breeding occurred at the Spring, Mile 4, and X89 sites; woodrats were also found regularly at the Rockslide, Stream, and Peaks sites. (See fig. 2 for general location of Sagehen Creek.)

andesite derived from Pliocene volcanic activity, interspersed with occasional andesitic breccia deposits (Burnett and Jennings, 1962). One bed is prominent along the northern side of the valley and can be traced as a series of outcrops through the forest from the Upper Spring site eastward to near the junction of the Sagehen road with the main highway. Andesitic breccia, although found elsewhere, is common along ridge tops as at the Peaks site and the uppermost portions of the X89 site. Pleistocene glaciation has also contributed to woodrat shelter by leaving beds of large glacial till in the valley bottom, particularly at the Lower Spring and Rockslide sites.

Weather in the Sagehen region is characterized by heavy snowfall from November through April, with heaviest fall from December through March. Summers have warm days and cold nights, and are generally dry, with occasional afternoon thundershowers for a few days of each month from May through October. Balgooyen (1976) published a detailed summary of weather prepared from station records.

One advantage of working at Sagehen was the information available from previous and concurrent work there on various mammals. Specimens and records indicated that bushy-tails had been collected there regularly since 1953. Many of my study sites were used previously by Nee (1967, 1969) in studying yellow-bellied marmots (*Marmota flaviventris*). While I was studying populations of woodrats in the rock outcrops, others were carrying on a variety of mammal studies in diverse habitats of the basin involving thousands of trap nights (Byrne, 1972; Reichart, 1972; unpublished studies of M. White, K. Delong, S. Smith, and others). In spite of that intensive trapping, only once was a bushy-tail captured away from either rocks or buildings, emphasizing the species' dependence on a highly specific habitat type.

Individual study sites are described in a later section of this chapter.

METHODS

From November 1970 through September 1974 I spent 155 days in the field studying bushy-tails. Most (129 days) were at Sagehen Creek Field Station in the summers of 1971-1974. Typically, I was at the field station at monthly intervals, June through September, for periods of 3-6 days, except during August when I remained for 18-25 days. The remaining field time was spent in three winter visits to Sagehen and in four collecting trips to other areas, primarily Modoc and Lassen counties in northeastern California. Other workers at the field station occasionally trapped bushy-tails in the course of their work or at my request. In June 1975 a final visit was made, but no trapping done.

In the first two summers, considerable effort was devoted to locating as much of the suitable rock habitat in the lower part of the Sagehen basin as possible. An area of about 21 km^2 (8 mi^2), from one side of the watershed to the other, and from above the Spring site to below the X89 site, was searched both on aerial photographs and by hiking surveys. The sites with evidence of extended woodrat use are indicated in capital letters on the map (figure 4). A few smaller rock areas showed signs of temporary use, but trapping there proved fruitless. An additional study site was located four miles (6.4 km) north of the basin (Mile 4) along State Highway 89, across the road from a campground on the Little Truckee River.

Most of the field data resulted from repeated live-trapping of marked populations, supplemented with radiotracking of certain populations for limited periods. Normal procedure for each visit to the field station was to check urine marks at the most accessible sites and set live traps that night at the most active ones, normally continuing trapping for three nights. In the following days, other sites were checked and traps set at other sites.

Home ranges were plotted on outline maps of the study sites by connecting the most extreme trapping sites or radio locations.

Urine Mark Survey

Field biologists for some years have recognized the white calcareous deposits on rocks as a useful sign of occupancy by N. cinerea and known that it is a urinary product (Bailey, 1898, 1918; Dixon, 1919; W. B. Davis, 1939). The white deposit results from the washing out, by rain or snow, of the organic components of the urine, for in caves it forms a dark thick layer (Finley, 1958; Nelson and Smith, 1976). Similar deposits in desert caves, probably from N. lepida, have been radiocarbon dated as several thousand years old; remains of plants collected by the earliest woodrat occupants have been used to trace climatic shifts in the southwestern U.S. (Wells and Jorgensen, 1964; Wells and Berger, 1967).

Although the exposed white marks of N. cinerea are probably not so persistent as those cave deposits, they certainly can remain for dozens of years, aided by the fact that subsequent generations add to and renew them. If an area is abandoned, and then reoccupied one or several years later, the new residents utilize the same sites and maintain them. The marks are useful in two ways. First, a rock outcrop can often be scanned from a distance to determine if it is or has been providing shelter to woodrats. The white edging on the rocks (plate 5a) is often emphasized by the presence of a bright red-orange lichen directly below it and occasionally a bright green one below that, both apparently depending on nitrogenous wastes washed down from the urine.

Secondly, urine marks, on close examination, can give a rough idea of the number of woodrats resident in an area. Fresh urine on the white mark leaves a shiny streak ranging in color from transparent yellow to opaque brown, is often sticky to the touch, and may have a musky or skunk-like odor. Heavy rains wash off much of the fresh urine from exposed marks, so it is helpful in interpreting the urine sign to know when it last rained. My experience demonstrated that even if only one rat is present, there will be at least a few places where fresh urine can be found. Age, sex, and length of residence seem to have no effect on whether or not the animal urine marks.

Invariably, if an area showed no fresh urine, setting traps was wasted effort. In a few cases I found areas with only one or two fresh marks, and trapping produced no rats. The fact that no more fresh marks appeared suggested that the marker was no longer present.

At one extreme, if only a few old deposits are freshly marked with single streaks in a limited area of the rocks, there is probably only one rat present (plate 5b). At the other extreme, an active breeding group with two or more females, an adult male, and a dozen or more young still resident is indicated when every white mark is covered with so many streaks in every direction that much of the white is obscured (plate 5c). With experience, an investigator can estimate the population level of a rock area by simply walking through it and looking closely at the urine marks.

Live-Trapping

Several types of live-traps were tried. Three types were found effective and were used together for most of the trappings. Two sizes of Havahart traps, 24 in. (61 cm) and 18 in. (46 cm) long, are sensitive and can catch even the smallest rodents. Their disadvantages are that they catch unwanted smaller species and are difficult to set in rough terrain. The third type, the 16-in. (41 cm) National or Tomahawk trap, is less sensitive than desirable, but can be set in more awkward places. Its folding version is useful for remote sites.

Two other types of trap were discontinued because they proved both harmful and less effective. The 10-in. (25 cm) Sherman trap (sheet metal) proved too small for N. cinerea, and seriously injured tails. A larger (19 in. or 48 cm) National trap has 1-in. (2.5 cm) mesh, in which woodrats can catch their heads while attempting to escape. Cranford (1970) had similar problems in using this trap for N. fuscipes. Its folding version also allowed some escapes.

Locations for individual traps were decided by intensity of fresh urine marking, visible presence of apparent dens, and prior trapping success at a given location. A regular grid would have been impractical because of both the uneven terrain and a limited number of traps. Woodrats are easily trapped (see results section), and populations in a given area were limited, so setting excess traps was unnecessary.

The most efficient use of traps, time, and effort dictated adjusting the number of traps in a given area for a particular trapping period. Normally 10-14 traps were set at one site. If a site was particularly active, or covered a large area, up to 20-25 traps might be set. Conversely, with only a few urine marks in a limited area, as few as 4-6 traps might be set.

Standard procedure was to set traps baited with rolled oats in each area for three nights at monthly intervals (more frequently in August and September). Traps were opened before dusk, inspected at least once between 10 p.m. and midnight, and again at closing time soon after dawn. In the early work, traps were left open 24 hours a day, but woodrats were rarely captured during the day, and the procedure was changed to avoid mortality from daytime heat in the rocks.

Some nights the schedule was interrupted by not opening the traps for a night or by closing them at the evening check. This was done to protect animals from adverse weather. Subfreezing temperatures can occur any night of the year there. Older bushy-tails are resistant to cold, but young under 4-5 weeks of age can die if left in a trap for several hours on a cold night. Evening rainstorms also interrupted trapping, for woodrats showed little resistance to wetting.

Handling of trapped rats was facilitated by a bag made of 8 mm (5/8 in.) nylon fishnetting. Animals were removed from the trap by placing the bag over the mouth of the trap and opening the door. Blowing sharply on its rump usually caused the rat to leap into the bag. Once in the bag, the animal could be restrained by holding it around the thorax from behind with a gloved hand.

Individuals were marked with numbered metal ("fingerling") eartags; toe clipping provided supplementary marking for the occasional animal which lost its eartag. Both operations were carried out easily by gently pulling ear and toes through the bag's mesh and seemed to cause the animals only momentary discomfort. On subsequent captures, the eartag number was usually read and recorded before removing the animal from the trap, since escapes sometimes occurred.

I weighed animals in the bag at each capture using a 500-gm Pesola scale, accurate to 1 gm. Sex, reproductive condition, pelage condition, and fresh wounds were also noted. Locations of capture within the smaller study sites were recorded either by reference to prominent landmarks or to simple sketch maps. For three of the largest study sites, where radiotracking was also done (Spring, Rockslide, and X89 sites), more detailed maps were prepared from aerial photographs, and capture locations were referred to a grid system superimposed on these maps.

Radiotracking

Live-trapping can provide only limited information about short-term movements of individuals, besides disrupting the normal pattern of those movements. Many questions about bushy-tails were left unanswered by trapping: the location of most dens and the identities of their occupants; the extent of home ranges and location of foraging areas outside the rock areas; degree of overlap or exclusion between home ranges; degree of interaction between individuals. These questions were among those I had hoped to answer by using radiotracking.

I built tracking transmitters following the design used by Cranford (1970) for tracking N. fuscipes. This transmitter design has been used with various modifications by many previous workers in tracking applications (Cochran and Lord, 1963; Verts, 1963; Mackay, 1970, p. 57). The transmitter is a crystal controlled blocking oscillator operating in the 27 MHz range ("Citizens' Band"). Those I built weighed slightly over 12 gm, operated for about 10 days on one mercury battery, and had an effective range of 100-300+ meters. I substituted a Motorola HEP 715 transistor in the circuit for the ones designated by Cranford because his were no longer available. Motorola subsequently replaced the HEP 715 with the HEP S0019, which also works well in the circuit. There were no other substantial design changes between Cranford's transmitters and mine.

After soldering, transmitters were coated with epoxy cement for mechanical strength. Waterproofing was considered unnecessary because woodrats are intolerant of wetting and were expected to keep themselves and the transmitters dry. Soldering of battery leads, wrapping with tape, and attachment of a plastic hospital ID bracelet for a collar were done at the field station just prior to putting the transmitter on an animal. Plate 6 shows a transmitter before and after these final steps.

Animals to be radiotracked were brought to the field station laboratory and anesthetized with ether. The copper antenna loop was slipped over the animal's head, and the plastic collar was snapped tightly and cut off short. Once the transmitter was attached, the animal was maintained in the station lab for 12-24 hours to recover and to make sure the collar neither choked the animal nor slipped off its head.

Previously, several animals in the laboratory at Berkeley had been tested with dummy transmitters for periods from several hours to several months. Although all animals initially were somewhat clumsy when released with transmitters, they were still able to run and climb well. Within a day or so, all accommodated to the load and recovered their normal agility. One male successfully bred in the lab while wearing a dummy transmitter.

Animals were tracked in the field using the receiver section of a modified 3-watt Citizens' Band walkie-talkie. A beat frequency oscillator (BFO) added to the receiver made the signal more noticeable and increased the useful range. The receiver could be switched among six different crystal-controlled frequencies. More transmitters than this could be followed in a single area because additional transmitters on the same channel could be built to pulse at different rates (60 per minute vs. 90 or 120 per minute), which were easy to separate by ear. A loop antenna (Cochran, 1966) and a signal strength meter wired to the receiver told the direction of the signal. By changing my position I could triangulate and locate the animal. If in doubt about a location, I usually took several additional "sightings."

BACKGROUND DATA

Capture Rates

Because of the patchy distribution of bushy-tails, and their readiness to enter traps, once a minimum number of traps were set, additional ones would have little effect on the number of captures. Thus, the common measure of captures per "trap-night" is considered inappropriate here. Instead I have calculated success in term of "area-nights," the number of nights multiplied by the number of discrete study sites trapped. In other words, if traps were set at the Rockslide for three nights, that would represent three area-nights. An unsuccessful area-night meant that no woodrats at all were captured that night in an area. Almost invariably these occurred when I set traps in spite of finding no fresh urine marks, or when no additional marks were found during or after trapping, indicating abandonment of an area.

In four summers of field work in the Sagehen basin and at Mile 4 (included with all Sagehen figures), a total of 146 area-nights was trapped, 123 of which were successful and 23 of which were not (table 11). In all, 107 individual woodrats were marked and released during that period in 428 captures. Seven of those individuals were captured in two different summers, and one in three summers, resulting in a total of 116 animal-summers.

Because analyses of the museum series previously discussed depend on the assumption that various sex and age classes are equally trappable,

TABLE 11
Live Trapping Effort for Neotoma cinerea at Sagehen Creek

Year	Successful Area-nights		Unsuccessful Area-nights		
1971	40		3		
1972[a]	31		8		
1973[a]	43		10		
1974	9		2		
	123	+	23	=	146

a. Radiotracking was also carried on in these years.

the live-trapping data were analyzed for this effect. For each area and trapping period "capture success" for each of four classes was calculated. The capture success, or capture/exposure index, was calculated by dividing, for each area, trapping period, and class, the number of captures by the number of nights and number of individuals present. Zero successes were not included in calculating the means, for in only a few cases were individuals known to be present but not captured during a given trapping period. These results (table 12) indicate that adult females are trapped more frequently per exposure, and statistics also suggest that this may be so (Mann-Whitney U test between adult males and adult females, one-tailed, corrected for ties, $p = .0681$.)

TABLE 12
Mean Capture Success for Neotoma cinerea at Sagehen Creek, 1971-1974

Sex	Age Class	Mean Capture Success	SD	N
Males	Adult	.838	.291	25
	First year	.813	.256	36
Females	Adult	1.142	.645	20
	First year	.858	.298	39
				120

That adult females also show much more variability in trapping success (higher standard deviation) suggests a further test. The above analysis includes recaptures within a single period, so "trap-happy" individuals can bias the live-trapping results considerably. A museum collector traps his animals only once, so a better comparison is made between the proportions of individuals captured on the first night of live-trapping in an area (table 13). "Exposures" represent the total number of first nights of trapping periods multiplied by the number of individuals of a class known to be present in each trapping period. Comparison on this basis gives a first night capture success of 84% for adult males and 87.5% for adult females, which are not significantly different. This result demonstrates that the difference seen in table 12 between adult females and other classes is most likely due to recaptures, so the museum results are unlikely to have been biased by differential trappability.

Adult females may be recaptured more frequently for two reasons. The energy demands of embryo growth and lactation could cause them to spend more time foraging, and they may also forage closer to home with young in the nest or while heavy with young, and thus be exposed more to traps set in the rocks. Cranford (1977) found that female _N. fuscipes_ significantly decreased their home ranges while breeding.

TABLE 13
Adult _Neotoma cinerea_ Captured on First Night of Trapping (when known to be present during that trapping period

Sex	First-night captures	First-night exposures	Ratio[a]
Males	21	25	.840[b]
Females	28	32	.875[b]

a. Ratio x 100 = percent of time an individual will be trapped on the first night when present

b. Fisher exact probability test, p = .275, not significant

Growth Curves

For a number of purposes, I needed to estimate the age of first year animals which I live-trapped. From the work of Finley (1958) and Egoscue (1962), it was evident that post-juvenal molt stages would be unreliable for aging. Use of total length measurements would have required anesthetizing each animal captured, and was thus impractical. Tests of hind foot length showed that the foot reaches adult size well before the rest of the animal. Weights, although having some drawbacks, could at least be taken easily, and proved more useful than originally expected.

Egoscue (1962) and Martin (1973) published growth curves for young *N. cinerea*, although Martin did not continue his curve to a sufficient age for my purposes. Egoscue's laboratory curves were based on only four individuals of *N. c. acraia*, which has already been found (museum section) to be smaller and less dimorphic than *N. c. alticola*, the subspecies which occurs at Sagehen.

Data for the curves presented here (figs. 6 and 7) were collected in three ways. For nestlings, data from laboratory born animals were used (open circles). Averages were taken of these laboratory values to avoid excessive weighting of the final curves.

The second set of data came from animals in the field whose birthdates were known (closed circles). In at least two instances, I captured a female just before and just after she gave birth. In one of these cases, where she was the only breeding female, I could estimate birthdates of all three of her litters by assuming a birth interval of 30 or 35 days. In the other case, presence of other females' young complicated matters, so I knew birthdates for only one litter. Because animals tended to lose weight with repeated captures in a trapping period, only one value, the heaviest (usually the first) is reported for a single trapping period; additional values are given for animals recaptured in later trapping periods.

The final set of values (open triangles) was derived less directly. Emigration, particularly of males, reduced the availability of older young with known birthdates. Other young of that age, however, were available either as immigrants or as residents whose birthdates were not so well known. To estimate ages in this third group, a linear regression for age and weight of known-age young was graphed and calculated (fig. 5). Birthdates for young of unknown age were estimated from that regression. Five females (estimated age at first capture, 55-63 days) and five males (66-81 days) were used; these estimated ages almost entirely overlap those of the known-age young from which the estimates were made. Only weights and estimated ages at subsequent trappings (fig. 5, open triangles) were used for further work (see fig. 5 for further description of method).

The initial estimates of age for this third group of young depend on the initial regression, and for that reason were not used further. On

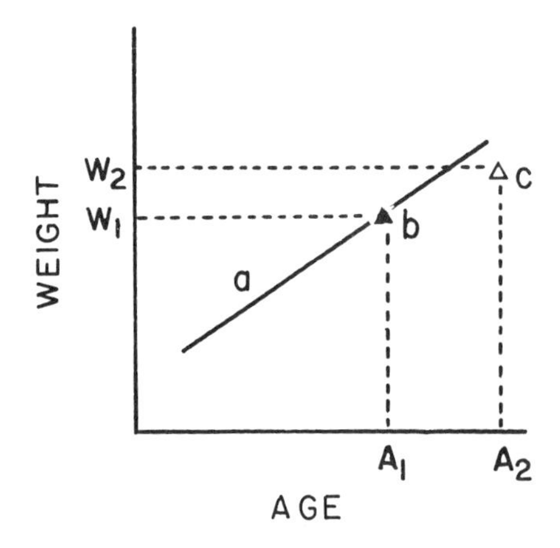

Figure 5. Method for estimating ages of older young for figs. 6 and 7 (diagrammatic).

a. Regression line is calculated for young of known birthdates (values represented as open and closed circles in fig. 6 and 7). (This regression is not identical to any regression shown in figs. 6 or 7 because it includes all data for young of known birthdates, not separated by age.)

b. Regression is used to estimate age (A_1) at first capture, based on initial capture weight (W_1) of young of unknown birthdate (but by molt stage, further growth, initial weight, and other evidence, clearly of the appropriate age range). This initial point (solid triangle) was not plotted or otherwise used in figs. 6 or 7, because it was derived from the first regression calculated.

c. At subsequent trapping, animal is heavier (W_2). Age (A_2) is estimated by interval between trappings plus original age estimate (A_1). Weight gain ($W_2 - W_1$) and interval ($A_2 - A_1$) are independent of original estimate. New point only (open triangle) is used in subsequent work (figs. 6 and 7).

the other hand, later values are less dependent on it because independent changes have occurred in both dimensions (age and weight), and we are primarily interested in that rate of change. In no case did subsequent points fall on the initial regression line. These less-dependent points were the only ones from these animals used in constructing figures 6 and 7.

These three sets of data were then plotted together for each sex. Theoretically we expect growth curves to be sigmoid, leveling off as the animal approaches adult weight. Many attempts were made to fit the present data to the three types of sigmoid curves described by Ricklefs (1967), as Martin did (1973) for his N. cinerea growth data. None of these curves fit satisfactorily, so the attempts were abandoned in favor of a simpler and more practical but less theoretically satisfying approach.

Inspection of the points for both sexes suggested that growth is nearly linear up to about day 40 and then slows down to another more or less linear rate which continues until at least day 140. (Egoscue's curves show a similar phenomenon.) Two linear regressions were then calculated for each sex, one for data before 40 days and one for data after 40 days (figs. 6 and 7).

The post-40-days regressions combine age-weight data from known-age young with growth rates of the estimated-age young which substituted for them (see above and fig. 5). These represent the best estimates available for growth rates in the absence of older known-age young.

Birthdates for other subadult young could then be determined by estimating ages at capture from the appropriate regression equation. For each individual, as many data as available or appropriate were used in making these estimates: weight at several capture dates, pelage condition, molt stage, time of year, health, and others. For further analysis here, estimated birthdates were grouped in two-week intervals (table 15 and fig. 20) to further minimize errors which may have been introduced by the method of estimating birthdates.

In bushy-tails, many young continue growing up to and beyond four months of age (present data; cf. Egoscue, 1962), which is well past the age of initial dispersal for most of them. At 140 days, many young approach or reach the theoretical asymptote of adult weight. Males grow

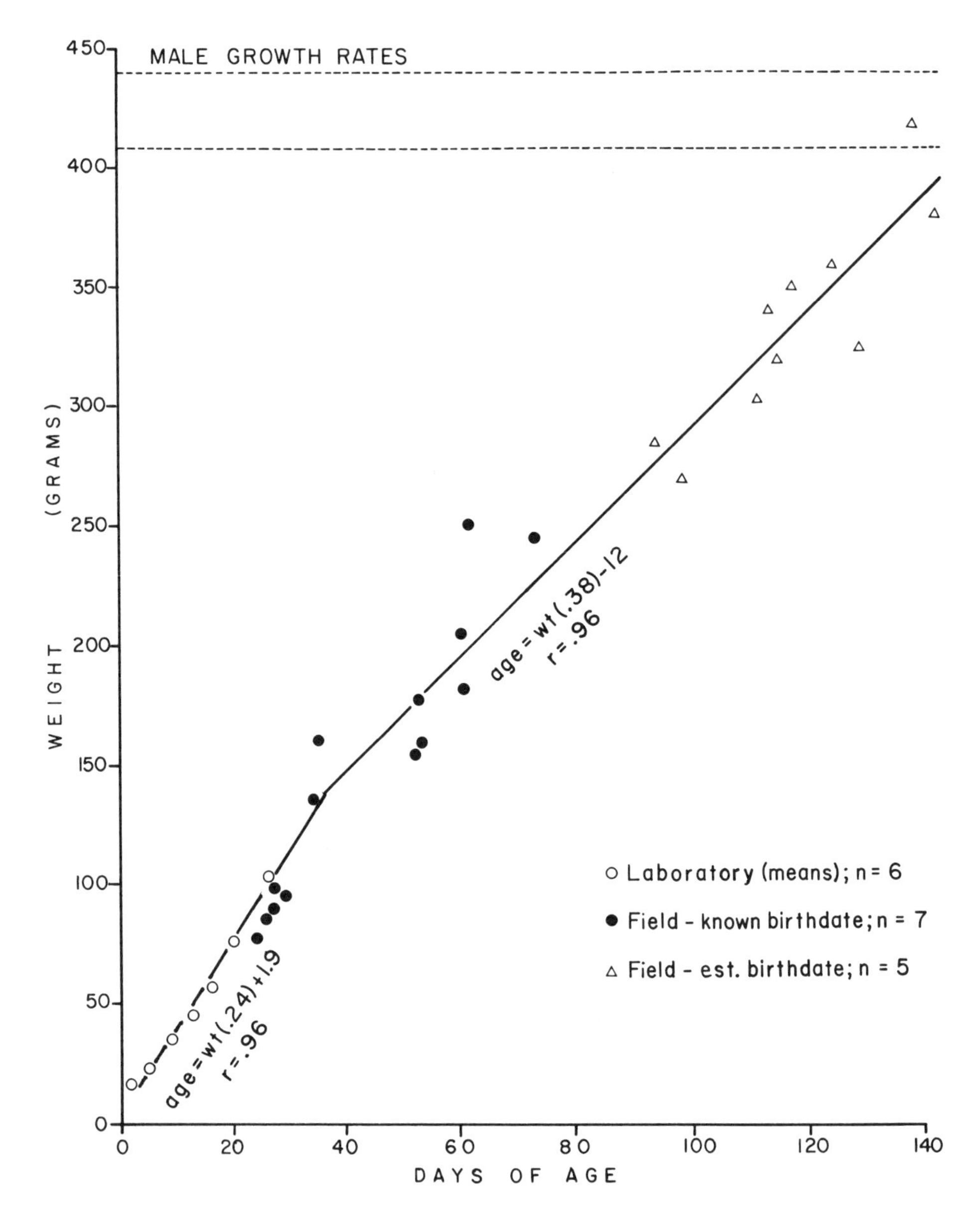

Figure 6. Growth rates of male <u>Neotoma cinerea alticola</u> (data combined from laboratory and field studies). Most laboratory values (open circles) represent means of up to six individuals; see text and fig. 5 for further explanation of data base. Horizontal dashed lines suggest range of theoretical asymptote (average maximum and minimum adult male weights at Sagehen, from table 8).

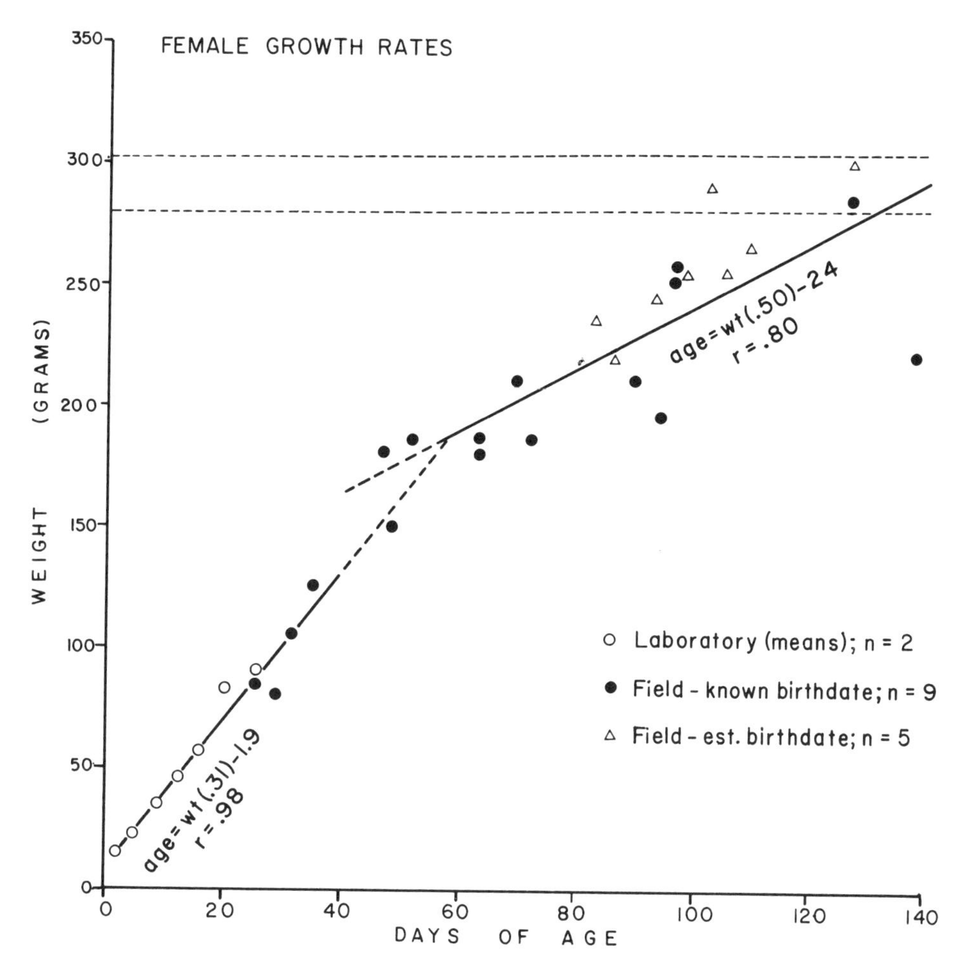

Figure 7. Growth rates of female Neotoma cinerea alticola (data combined from laboratory and field studies). Certain of the laboratory values (open circles) represent means of two individuals; see text and fig. 5 for further explanation of data base. Horizontal dashed lines suggest range of theoretical asymptote (average maximum and minimum adult female weights at Sagehen, from table 8).

at a faster rate (noted also by Egoscue, 1962, and Martin, 1973), which allows them to reach their greater adult weight almost as soon as females. For most young, this means their growth continues into the fall. That may be important, because large size apparently aids in winter survival of woodrats (cf. Brown and Lee, 1969).

STUDY SITE SUMMARIES

Following are brief descriptions of each study site (see fig. 4 for locations in Sagehen Creek area) and outlines of changing patterns of occupancy in each over the four years of my field work. Practically all field data resulted from these long-term records, and are best understood in that context. Birthdates were determined from the regression curves and other data as described above. Assuming a gestation period of 30 or 35 days (see laboratory chapter) also aided in dating successive litters, as did regular assessments of breeding condition for adult females. Immigrants to an area already containing breeding animals and their young could usually be identified because they did not fit the age pattern of locally born young.

Scientific names of plants mentioned are listed in table 14.

Spring Sites

There are two major rock areas about 150 m apart at this location. The Upper Spring site is an outcrop of andesite with two prominent elevations and a steep cliff along the center (fig. 8 and plate 7a and b). The rock has fractured into flat plates up to 2-3 m across, and most of the rock area shown in figure 8 below the "peaks" is an extensive and relatively unstable talus slope (rock area = 0.6 ha). Aspen covers much of the rock area, with cream bush, serviceberry, and bitter cherry in places. The forest surrounding the outcrop consists primarily of lodgepole pine with some Jeffrey pine and white fir. A tent platform belonging to the field station, and an open dry meadow lie between the Upper and Lower Spring sites.

The Lower Spring site (fig. 9) is an extensive (0.3 ha), nearly level, deposit of glacial till (plate 7c) with a fringe of aspen on its uphill side; on its stream side is a strip of riparian vegetation, primarily of willow with mountain alder, creek dogwood, and a small patch

TABLE 14
Scientific Names of Plant Species at Sagehen Used in Text

Scientific name	Common name
Abies concolor	White fir
Abies magnifica	Red fir
Pinus jeffreyi	Jeffrey pine
Pinus murrayana	Lodgepole pine
Pinus ponderosa	Yellow pine
Juniperus occidentalis	Sierra juniper
Arctostaphylos patula	Greenleaf manzanita
Ribes aureum	Golden currant
Holodiscus discolor	Cream bush
Purshia tridentata	Bitter brush
Rosa sp.	Wild rose
Prunus emarginata	Bitter cherry
Amelanchier alnifolia	Serviceberry
Alnus tenuifolia	Mountain alder
Castanopsis sempervirens	Chinquapin
Populus tremuloides	Aspen
Salix sp.	Willow
Rhamnus alnifolia	Alderleaf coffeeberry
Ceanothus velutinus	Tobacco brush
Ceanothus prostratus	Squaw carpet
Cornus stolonifera	Creek dogwood
Chrysothamnus nauseosus	Rabbit brush
Artemisia tridentata	Sagebrush
Allium sp.	Wild onion

of alderleaf coffeeberry. Across Sagehen Creek from these sites is a smaller rock outcrop, mostly of small rocks; a few larger boulders at its base occasionally harbor woodrats.

In 1971 a single pair bred in the Upper Spring area, producing three litters born May 17, June 21, and July 20. The adult female was last trapped July 19 and disappeared about August 10-11 for—although she was not trapped August 12-14—two young which she must have raised to weaning age were trapped on the 13th and 20th.

No fresh signs or woodrats were found in the Lower Spring area until late August when two of the young females, born in May, moved there. The adult male was also captured there periodically as well as at the Upper Spring.

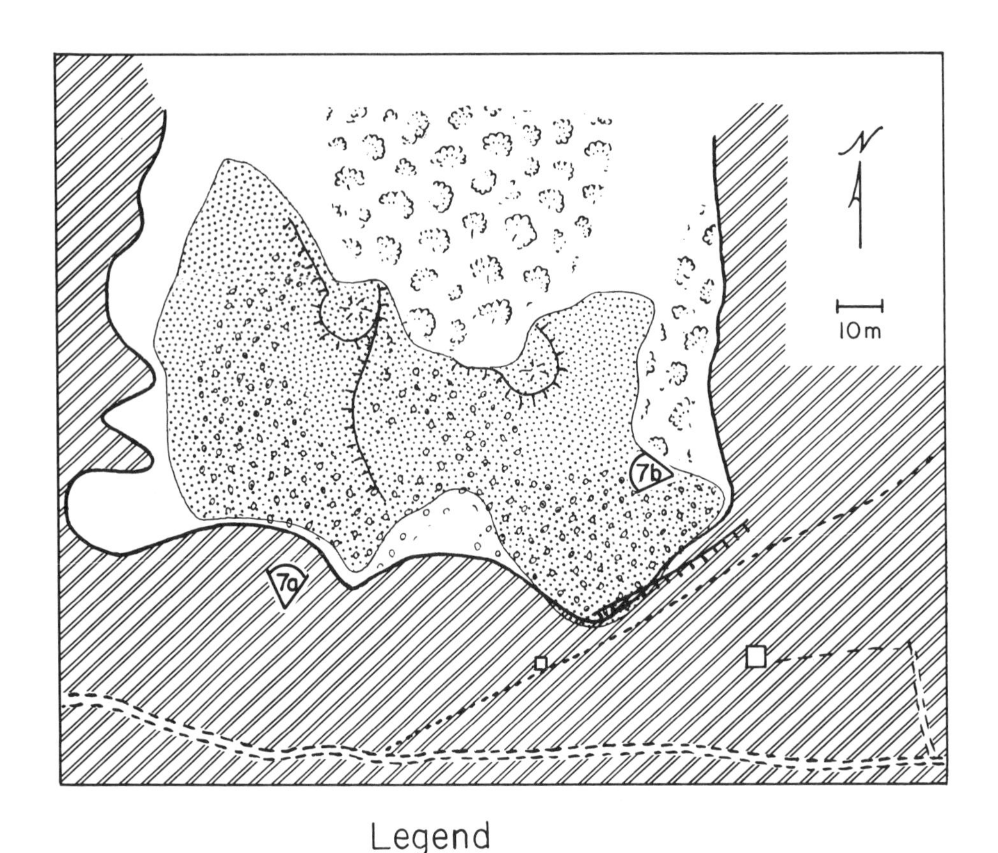

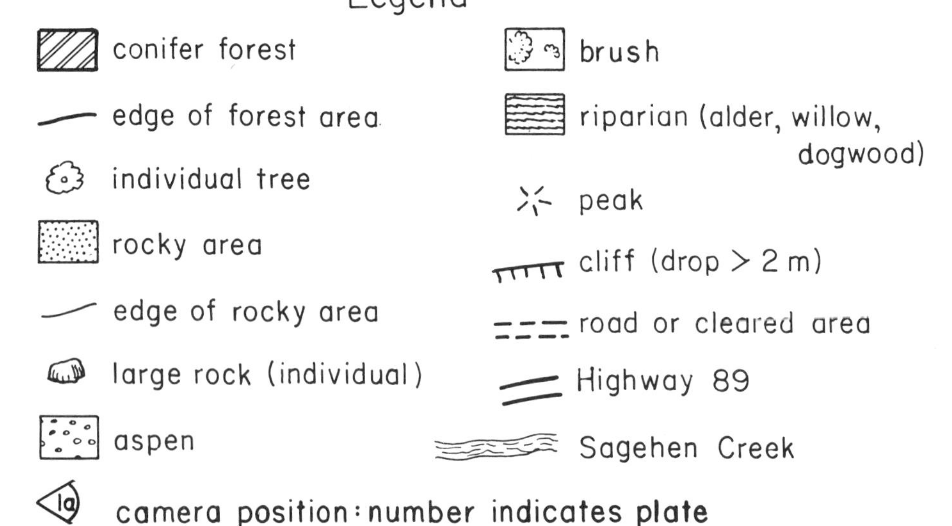

Figure 8. Map of Upper Spring study site. (Legend applies to figs. 8, 9, 14, 15, and 16, so not all symbols are found on this map.)

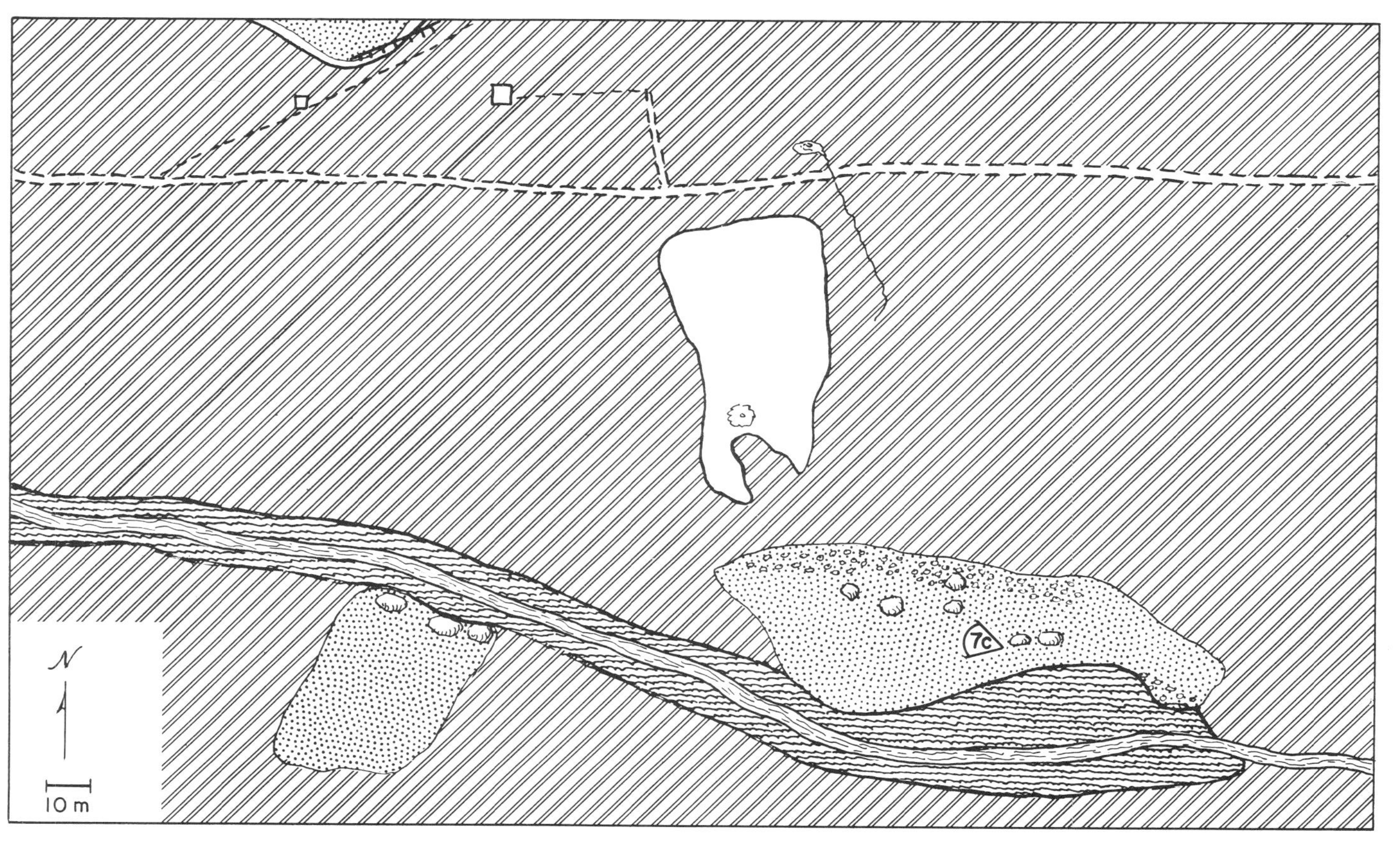

Figure 9. Map of Lower Spring study site. Edge of Upper Spring study site is seen at top of map, indicating their relationship. (See fig. 8 for legend.)

Two immigrants appeared and settled at the Upper Spring: a female (1600) in mid-August, and a male (1108) at the end of August (fig. 10). The female's home range overlapped those of the local animals, while the male was trapped only on the periphery.

At the last trapping in late September, the two young females were still at the Lower Spring. At the Upper Spring were the adult male, the two immigrants, and another subadult female born there July 20.

In 1972 I did not trap until July. At that time I found three breeding bushy-tails: the immigrant male (1108) and two of the females

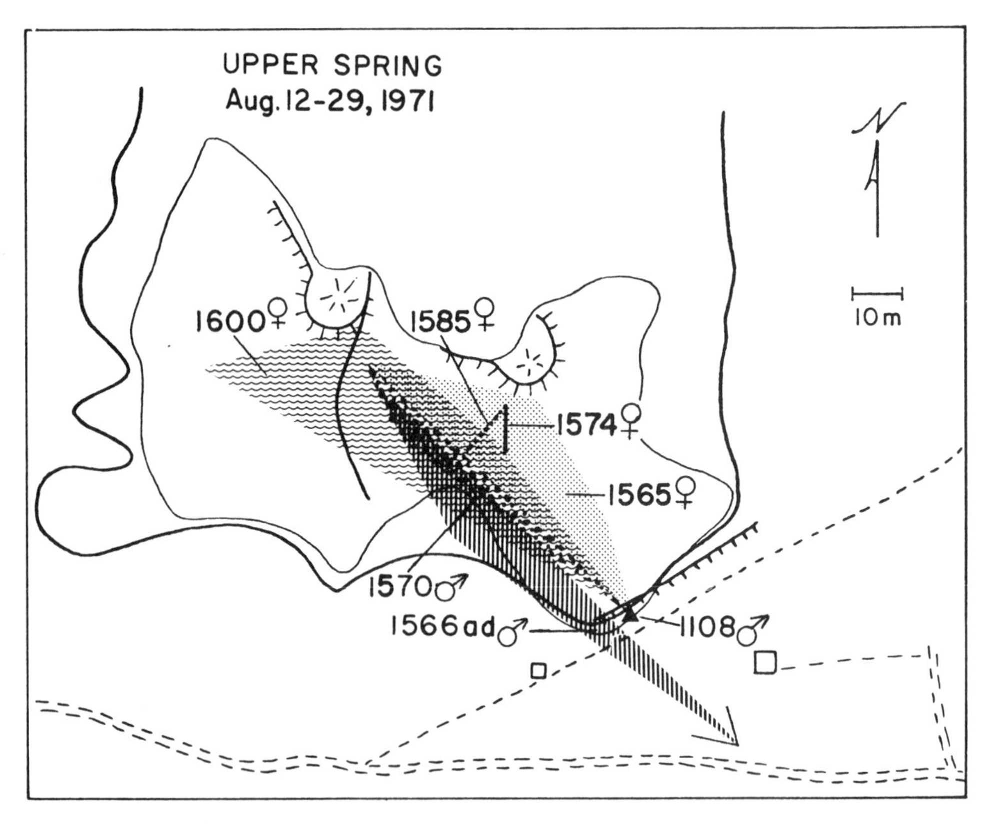

Figure 10. Home ranges at Upper Spring, based on live-trapping, Aug. 12-29, 1971. Numbers next to sex symbols give eartag numbers of individuals; except as otherwise designated, all are young of the year. Shaded areas indicate home ranges for designated animals. Solid and dotted lines are used for home ranges of animals (1574, 1585) caught at only two places; triangle indicates single capture site. 1566 was also captured at Lower Spring (arrow).

born there the previous summer, one in the Upper and one in the Lower area, where they had been in the fall. Neither female was trapped in the other area that summer, and the male was trapped only in the Upper area.

Data are less clear concerning litters: the Upper Spring female had litters May 20, June 20, and probably July 20. The Lower Spring female had litters May 5 and June 3. Since she was still lactating on July 22, she probably had another litter, but only one young was captured of an appropriate age, at the Upper Spring, and it was assigned to the other female. Because litter dates are unclear, the question of immigrants is also unclear, although there appear to have been none.

In late September, animals (all except 1130) at both areas were radiotracked. Home ranges at the Upper Spring overlapped extensively, and all but two rats had separate dens (fig. 11). The adult male (1108) and subadult female 1118 (born May 20) apparently shared a den. Subadult male 1130, born July 20, did not receive a radio because I had too few transmitters; without a radio, his den could not be identified. Male 1120, born June 20, was nesting in a hole in a pine tree, 8-10 m above the ground, next to the vacant tent platform. Besides his not nesting in the rocks, his home range only barely met the others. He was later trapped on October 14 in a woodpile at the field station (0.4 km away), where a woodrat—presumably him—had been seen about two weeks previously. Radiotracking at the Lower Spring (no figure) involved the other adult female and three more subadult females. Each maintained a separate den under a large boulder, and had overlapping but distinct home ranges. Since there were at least eight woodrats left at the Spring site in the fall of 1972 (of the ten radiotracked, one (1120) was known to have left, and another died in a trap accident at the end of the radiotracking), I was surprised in early June of 1973 to find no fresh sign at the Upper Spring, and only a few spots at the Lower. Trapping produced only a single adult male (1598) on June 6 at the Lower Spring. This male had been captured as an adult in 1971 and 1972 at the Stream site, about 0.4 km away, and this capture at the Spring was the last time he was seen. The Spring area again was completely deserted in July. Four subadults appeared in early August and were regularly trapped through late October (figs. 12 and 13). Their weight at first

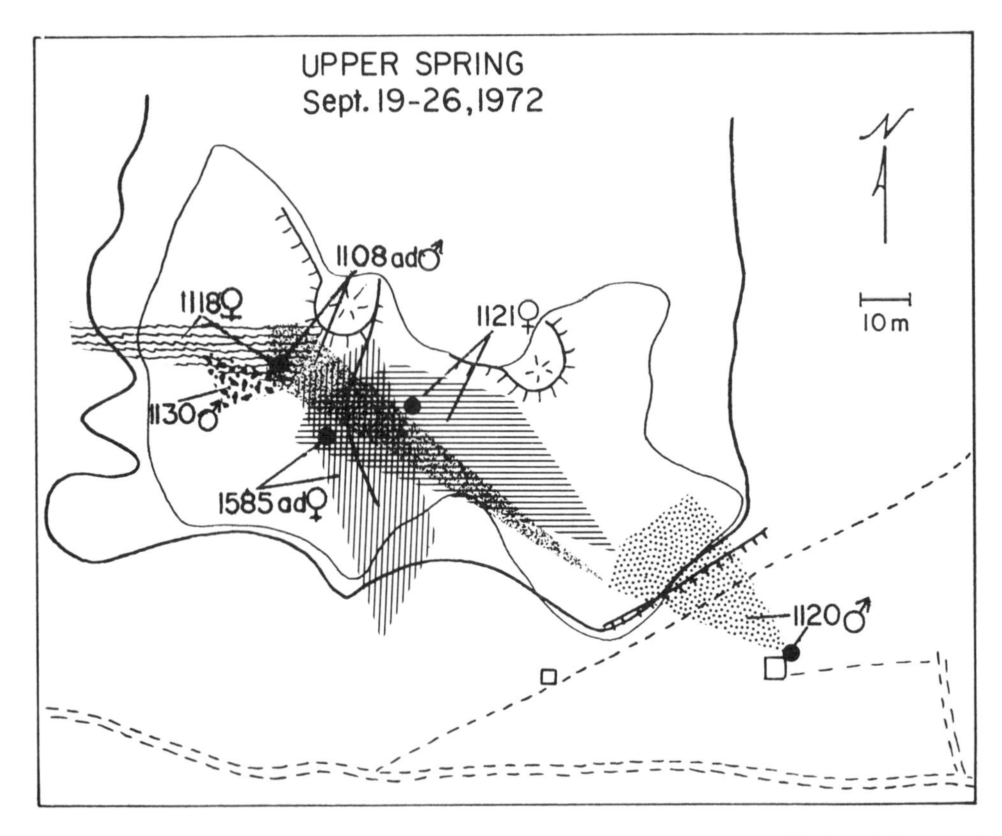

Figure 11. Home ranges at Upper Spring based on radiotracking and live-trapping, Sept. 19-26, 1972. Solid circles indicate den sites; other symbols as for fig. 10. Data for 1130 based on trapping only, so no den site is shown.

capture suggested that all were probably born May 28-31. A single male (1176), born about June 12, was captured across the stream only once, in September, and not seen again.

The Spring area was again deserted by June 1974. No fresh signs could be found in July, August, or September checks, nor were any present in June 1975.

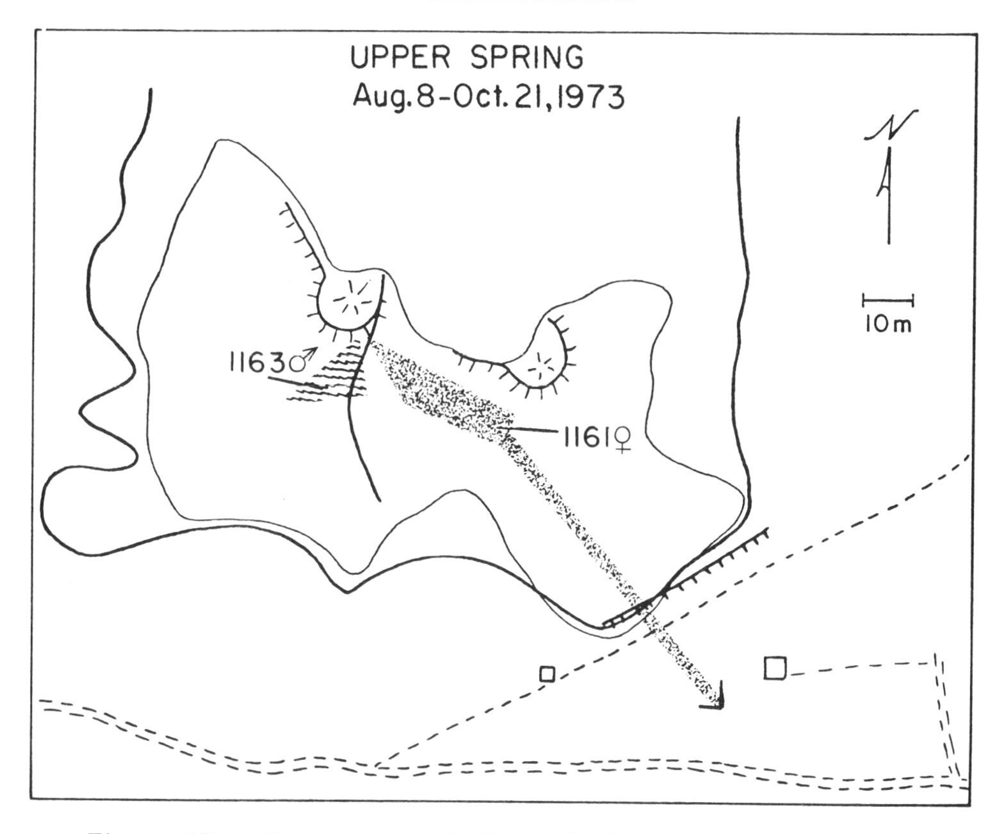

Figure 12. Home ranges at Upper Spring based on trapping, Aug. 8–Oct. 21, 1973. (See also fig. 13; symbols as for fig. 10.)

Rockslide Site

This nearly level deposit of large blocks (many more than 2-3 m in diameter) of glacial till (fig. 14; plate 8a) was designated the "Rockslide" by Nee (1967). The rock area is extensive (0.7 ha), and white urine deposits may be found in numerous sheltered places throughout. Aspen fringed both the stream and uphill sides of the area until most aspen on the stream side was cut by beavers in 1973-74. Wild rose also

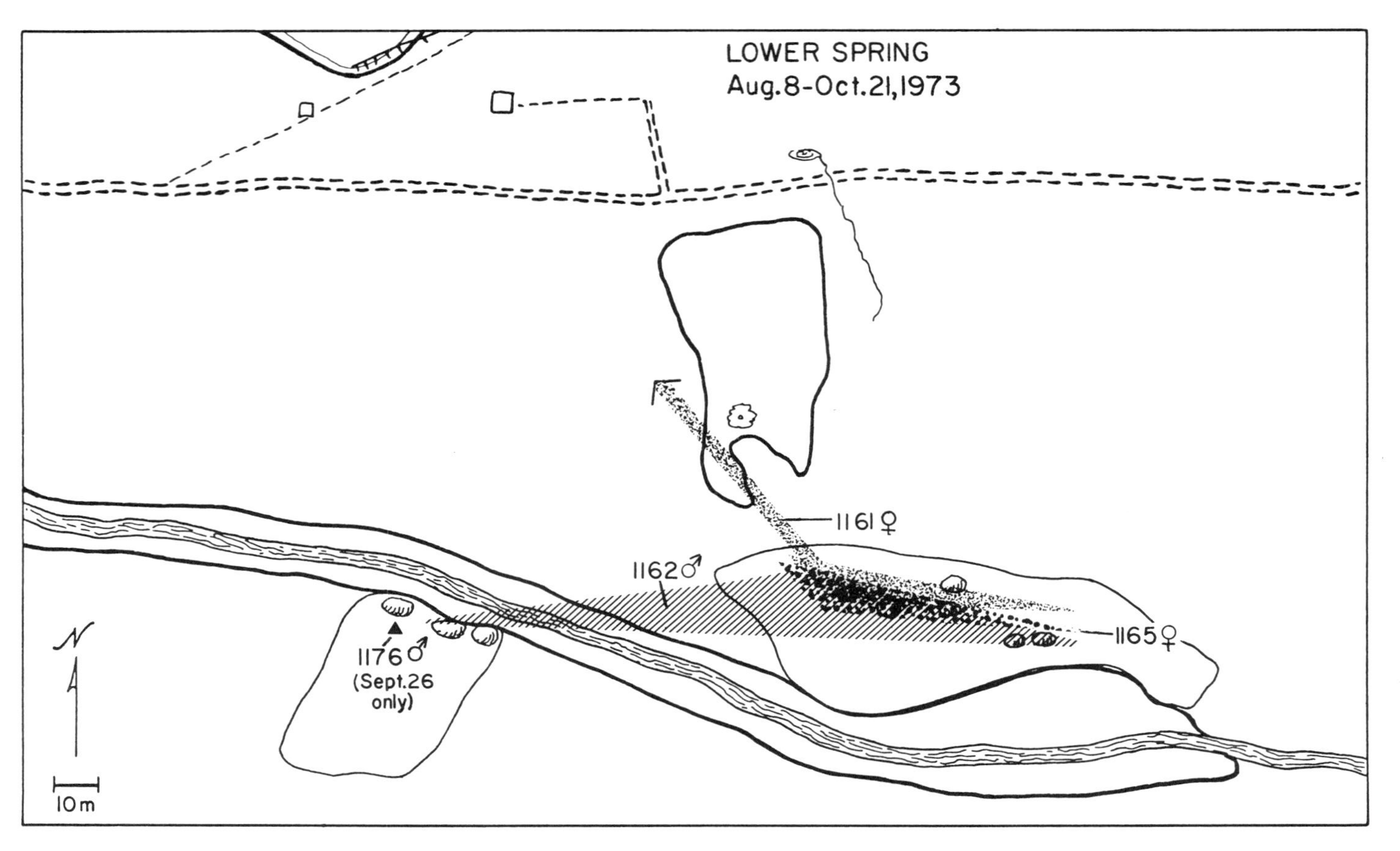

Figure 13. Home ranges at Lower Spring, based on live-trapping, Aug. 8–Oct. 21, 1973. (See also fig. 12; symbols as for fig. 10.)

grows along the stream side. Yellow pine forest surrounds the area, with small brush fields between forest and rocks on the north and west edges (greenleaf manzanita, bitterbrush, tobacco brush).

The Rockslide, judging from the extent of old urine deposits, frequently housed breeding groups of bushy-tails in the past. No breeding occurred there during the present study, although the area was frequently recolonized. In October 1970, just before this study began, a sub-

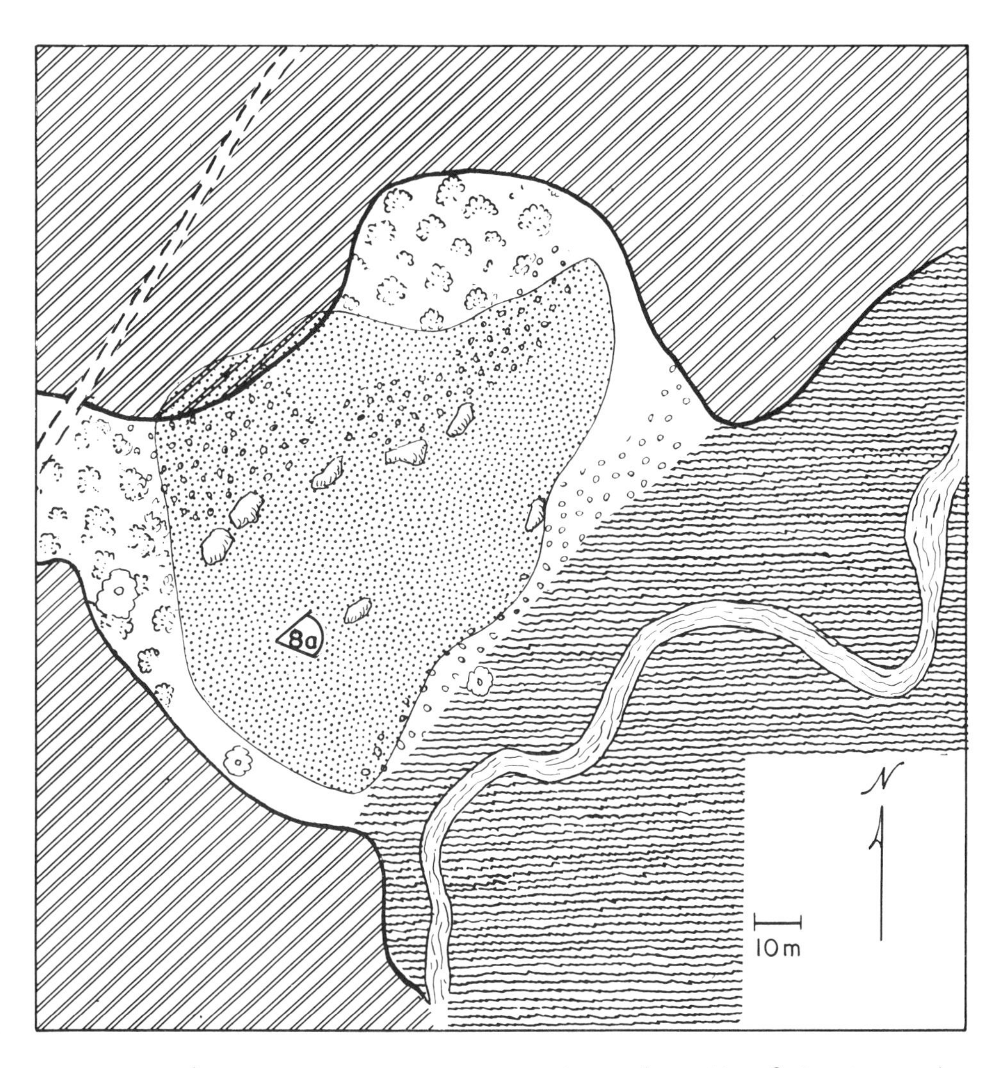

Figure 14. Map of Rockslide study site. (See fig. 8 for legend.)

adult male and subadult female were collected from the Rockslide by a mammalogy class; in 1971 no woodrats were there again until mid-August, when a single subadult male settled there. He remained through late September, but disappeared during the winter.

On July 21, 1972, I found three new immigrants, a male and two females, which had been born between April 25 and May 7. As a first test of the radiotracking equipment, the two females carried transmitters August 28-September 5. They had dens about 75 m apart at opposite ends of the Rockslide, and completely distinct home ranges (no map given). The male during this period was trapped over the whole area, and his range overlapped those of both females. In mid-October, a visiting student who was unaware of my study collected the male and one of the females; the other female was never seen again. (If these two specimens had been considered in the museum section of this report, they would have provided a conservative element to estimates of polygamy.)

In June, July, and August of 1973 the Rockslide was again deserted. On August 24-25 I live-trapped three subadults 12 and 19 km north of Sagehen (2 males weighing 300 and 365 gm, 1 female weighing 220 gm), and released them with transmitters into the vacant Rockslide on August 26. (These individuals are not included in the counts previously given of woodrats trapped at Sagehen.) My intent was to simulate immigrants just arrived at a new home area, and I wanted to observe the establishment process. Throughout the entire observation period (11 days), all three animals slept in a different place nearly every day, and daytime movements were observed several times for both males (no map given).

Both of these behaviors contrast with all other radio observations I made of "established" woodrats, which use the same den consistently and almost never move about in daytime. The males' ranges overlapped through the experimental period, and they even alternated in using the same daytime resting places. No clear pattern could be established except that they were avoiding each other. In the last few days they tended to use opposite ends of the rock area.

The female moved throughout the area, her range overlapping that of both males. On the last two days of tracking she used a den between those of the males.

Although Fitch and Rainey (1956) found such increased movements and use of several dens typical of reduced densities in *N. floridana*, in the present case the density was identical to that of the previous year in the same area, when the residents exhibited more typical patterns of movement and den use.

Radios were removed September 6, and by September 29 only the female and the larger male were present. By October 19-21, only one male was left, and he was not there the following June (1974).

No colonization took place in 1974, nor was there any fresh sign in June 1975.

Across 89 ("X89") Site

Across State Highway 89 from the road to the field station stands a small hill ("6665" on USGS 15' Truckee quadrangle, 1955 edition). At the crest of the hill are two outcrops containing a few crevices and some talus (fig. 15). After the 1960 fire removed most vegetation, the Forest Service terraced the steep slope with bulldozers (records of U.S. Forest Service, Tahoe National Forest, Truckee Ranger Station), exposing in the process several areas of large boulders near the base of the hill (plate 8b). Most "roads" shown in fig. 15 are old bulldozer paths and impassable. Since the fire, the area has developed into open brush, consisting of about equal parts of bitter cherry, greenleaf manzanita, bitterbrush, golden currant, and tobacco brush. Sandy areas are dominated by rabbit brush, and many of the rock areas support bitter cherry and manzanita. Scattered stands of yellow pine and white fir remain from the original forest; seedling pines have been planted, but are inconspicuous.

Trapping did not begin at X89 until 1972. A brief check in August 1971 revealed a small amount of fresh urine marking in the upper area, but no traps were set; I was then unaware of the rocks at the bottom of the hill because they were hidden from routes I took. A year later (August 1972) I discovered these boulders (plate 8c; fig. 16), and found extensive fresh urine on the rocks, although there were no old white marks. Trapping August 28-September 1 in the lower area produced an adult male and four subadult females, whose birthdates were estimated as May 5, 7, 16, and 24.

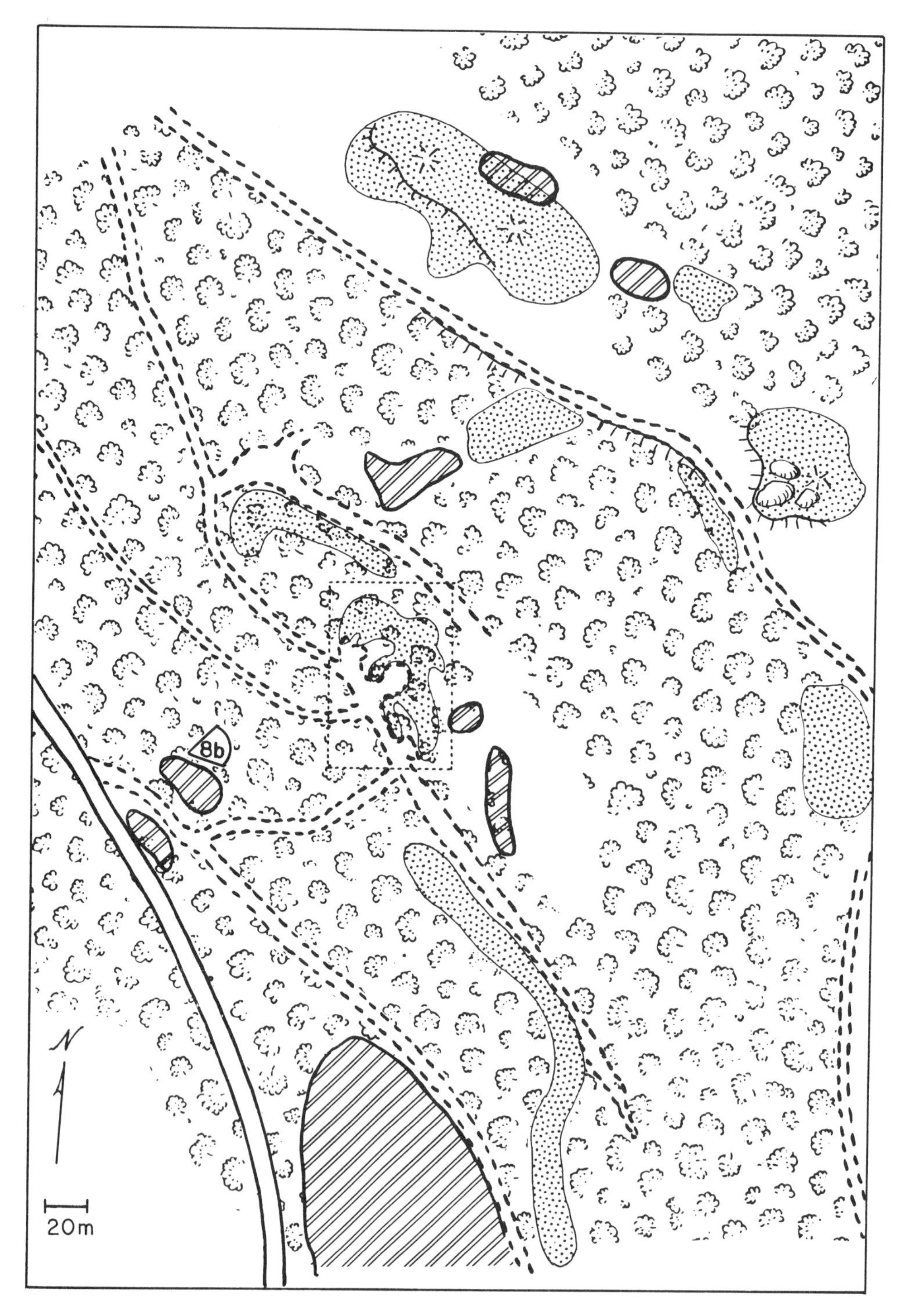

Figure 15. Map of X89 study site. (See fig. 8 for legend.) Rectangle indicates location of den sites (figs. 16 and 18).

Returning the first week of June 1973, I found the male had been replaced by a different male, and the smallest of the four females had disappeared, but the three remaining females had all bred. Fourteen young apparently born there were captured June through August (7 males, 7 females); an additional subadult male, born about June 11, was presumed an immigrant because he was not captured until September 29, and then only at the top of the hill. One female was known to have given birth on June 29. Other litters were estimated as born May 5, 11, and 14, and June 5 and 16, but assignment to particular females was not

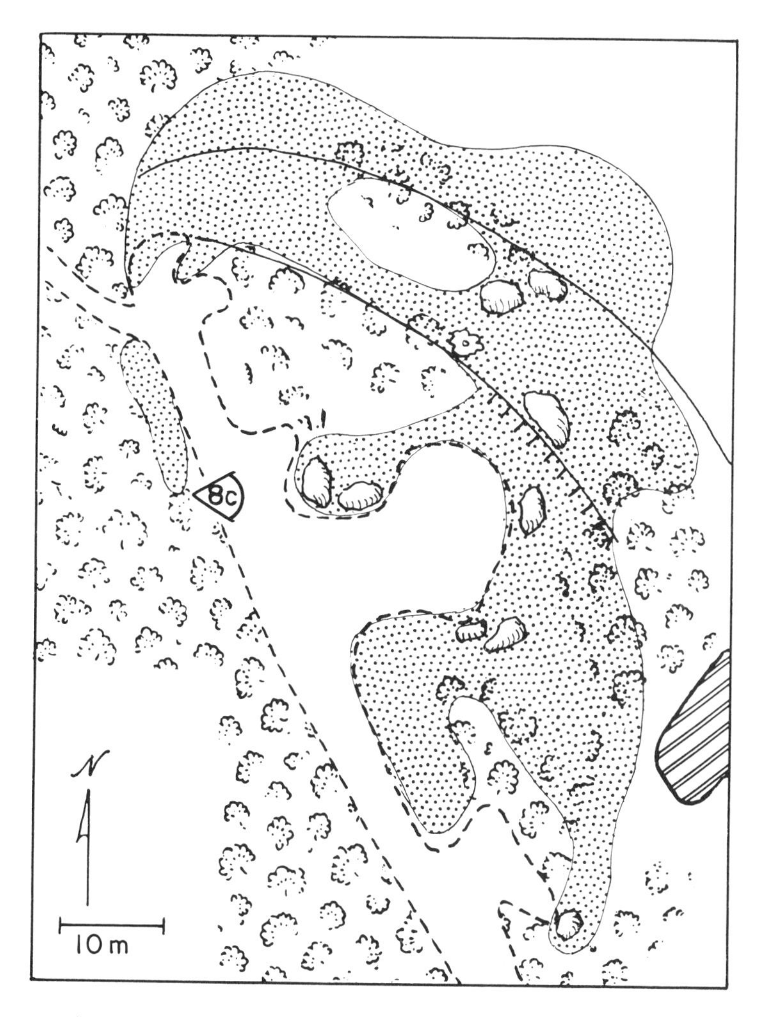

Figure 16. Map of central (den) area of X89 study site. (See fig. 8 for legend.)

possible. By August 14-15, only the four adults and five of the subadults remained in the area: three females in the central area and a male and a female (1160 and 1167) at the top of the hill (see fig. 17).

From August 16 to September 6, the animals living in the central area were radiotracked, except one subadult female, which died in a trap. A total of 273 radio "fixes" and 25 captures was recorded during that period for the six animals. (They were trapped during the middle of the period to change batteries.) The two subadults at the top of the hill were last trapped on August 18 and not seen again. The ranges (fig. 17) have imprecise boundaries because of the difficulty in exactly locating moving animals, particularly at greater distances. The range for 1159 is truncated because her radio worked for only a few days and was so weak that she got out of range by moving up the hill.

Several interesting points emerge from the radiotracking. Within the rock area (fig. 18), the rats' ranges overlap considerably. Several times, based on radio fixes, individuals appeared to visit dens other than their own, both with and without the resident at home. The adult male (1147) in particular apparently visited the dens of all but female 1159. Females 1134 and 1132 both appeared to visit the male's den as well as spending time near him (out of my visual range) away from the dens, but in the rocks. On one occasion, 1147 and 1132 were together for nearly an hour just before dawn about halfway between their houses. After they had been together for 35 minutes, a long-tailed weasel (*Mustela frenata*) passed within a meter of them, but continued on without investigating. Subadult 1166 also made several visits to the den of 1132; since their foraging ranges overlapped considerably, possibly 1166 was the offspring of 1132.

The arrangement of the dens is intriguing, although the significance is not clear. The adult male had the most central of the six dens, the adult females the next most central, and the two subadults the most peripheral (fig. 18). Examination of the den sites in the field did not show obvious differences in quality between them, for all were well under large boulders (see plate 8c). Actual construction by woodrats could be seen at only two dens, those of 1147 and 1132; the other dens were either better protected or hidden or had less effort put into their construction.

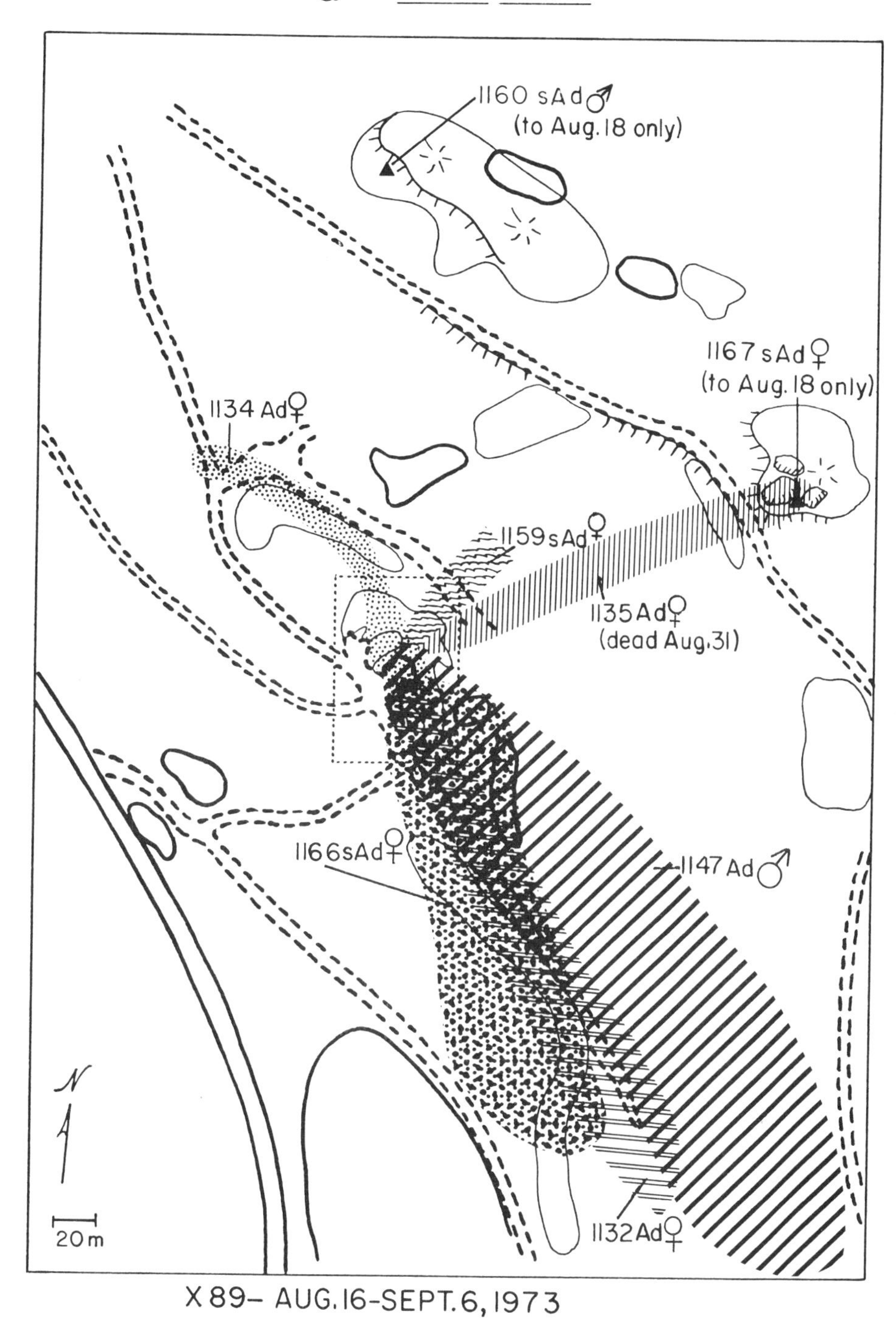

Figure 17. Home ranges of Neotoma cinerea at X89 study site, Aug. 16-Sept. 6, 1973, based on radiotracking and live-trapping data. (See also fig. 18.)

Comparison of the den area (fig. 18) with the foraging ranges (fig. 17) is revealing. Although the animals denned close to each other and apparently visited each other's houses, the foraging ranges of the three adult females were completely separate, fanning out in opposite directions from the center. The two subadults' ranges more or less overlapped ranges of the adult females they lived closest to. The adult male covered the most ground, moving as much as 350 m from the rocks and back within a matter of minutes. Not shown on the map is that he was also trapped in the upper area, but not during the designated period.

The ranges are nearly linear because these animals commonly moved out to the end of their ranges and back several times a night, sometimes spending as much as an hour or two away from the den, and other times making the round trip in as little as five or ten minutes. Exact paths taken could not always be determined because of the difficulty in locating a rapidly moving animal. In general, animals avoid open areas, staying under heavy brush or among rocks. Different routes might be taken going out and returning on the same trip.

Berries were available on the abundant bitter cherry, manzanita, and bitterbrush during this period. Although unable directly to observe the rats feeding, I presume these plants were an objective of their forays. I could see little difference between bushes close to home and those the rats went to. Cranford (1970, 1977) found that areas to which N. fuscipes traveled regularly were those containing seasonally abundant foods. The only storage I could find, at 1132's den, consisted almost entirely of green leaves of bitter cherry, which had obviously been clipped only a few meters away within the rock area, and contained few fruits.

The disappearance of 1135 during the radiotracking period was the only well-documented case of predation during the study. She had been the last female to give birth in the area (June 29), and probably the only one to have had three litters there that summer. She apparently was subordinate to the others in the area because even in early June she had a number of wounds on her rump, as seen in lab animals under repeated attack. She was last observed by radio in her den at 8:40 p.m. on August 27. Attempts to locate her later that evening were unsuccessful.

No more signals were heard from her transmitter until the evening of the 31st when a weak signal was followed to a 7 cm diameter hole in the

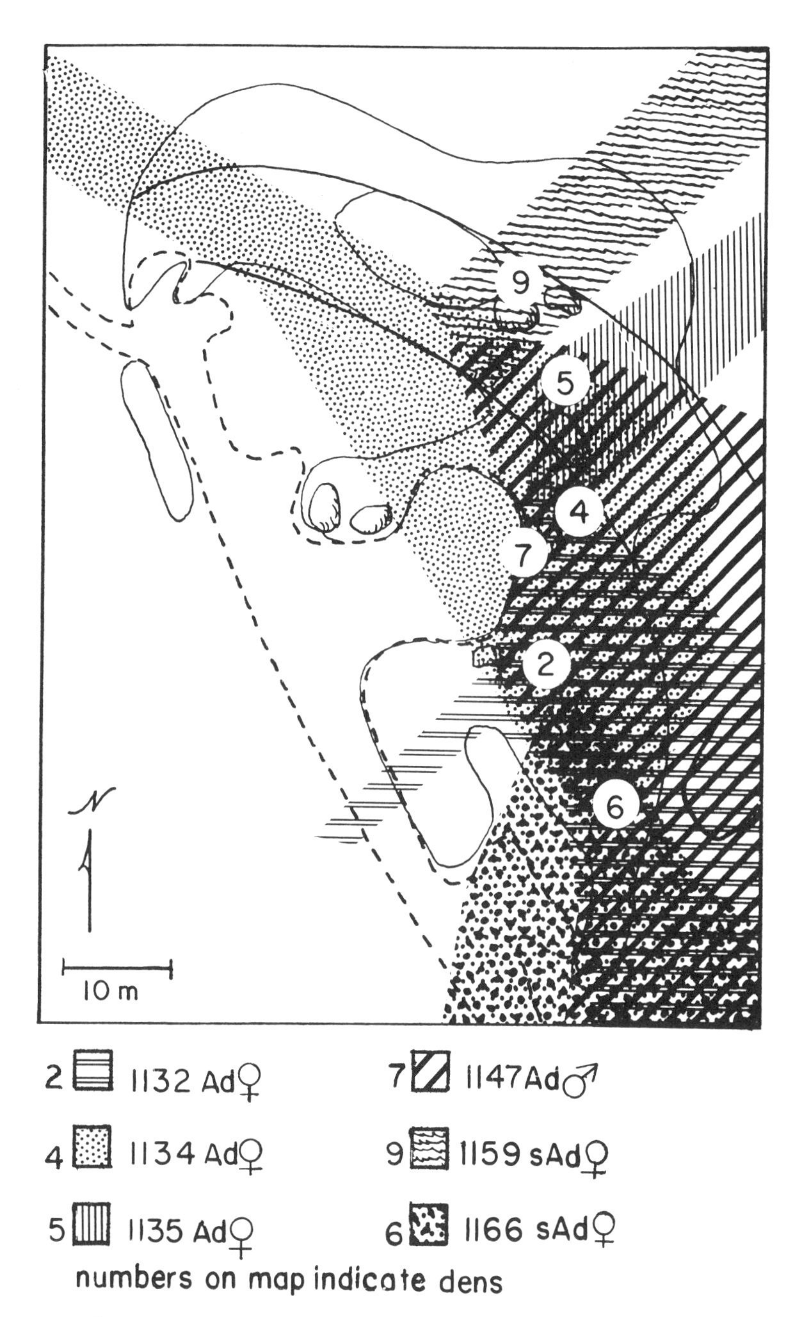

Figure 18. Den sites at X89 study site, Aug. 16-Sept. 6, 1973 (detail for fig. 17; see also plate 8c).

ground about 350 m WNW of the rocks, across the highway and out in a brush field. Next morning, about half a meter below the surface, I found the transmitter with her remains. The back of her skull was gone, typical of a weasel kill, and all else that remained were the inside out skin, the feet, tail, and part of the intestines. Because of the bite to the back of the head, the size of the hole in the ground, and the sighting of a weasel at the study site four days before she disappeared, there is little doubt of the predator.

Radios were removed on September 6. At the end of September all of the remaining woodrats were still present, with the addition of an apparent immigrant (subadult male) at the top of the hill. The same arrangement still held in mid-October.

Returning early in June 1974, I found that a Forest Service crew had removed a number of boulders from the center of the den area to block off the road leading into the site. The equipment tracks were fresh. A urine check revealed that there had probably been few woodrats present before the disturbance. Trapping produced only 1134, not in breeding condition. By late July she had been joined in the main area by an immigrant subadult male. A female subadult of about the same age had moved into the upper rocks. All three were in the same places in late September.

Stream Site

This deposit of glacial till is located inside a bend in the intermittent stream flowing into Sagehen Creek just above the field station. The Stream site never housed a breeding group during the study. An adult male (1598) was first captured here in 1971, recaptured in 1972, and last seen (once) at the Lower Spring in 1973. In 1971 three immigrant young were captured once each: two in August, one in September. In 1972 a young female was captured three times in August. None of these subadults had been tagged previously, nor was any one of them collected later. No fresh signs or woodrats were found at the Stream site in 1973 or 1974.

Peaks Sites

Located at the top of the ridge almost directly behind the field station are two rock outcrops about 300 m apart (designated East and

West). Each is only 50-100 m in diameter, and neither supported a breeding group during the study. They regularly served as a way station for dispersing first-year woodrats. In 1971, no rats were found, but old urine deposits attested to prior occupation.

Trapping September 3-5, 1972, I caught a subadult male and two subadult females at East Peak, and a subadult female with a 35-day old male (1144) at West Peak. Ten days later, 1144 was caught by a kill trap in a tent at the field station, 2.2 km away.

No woodrats survived the winter at the Peaks, but on July 3-4, 1973, a young female (1150) that had been tagged and last seen on June 4 at the X89 site was trapped at West Peak; her birthdate was estimated as May 4. Thus, within 28 days, while between one and two months of age, she had traveled 3.2 km. She did not appear in subsequent trapping. Three more subadults were trapped at the Peaks that summer and fall, each alone and in a single trapping period only. No fresh signs or woodrats were found in 1974.

Mile 4 Sites

Driving north toward Sierraville from Sagehen, one passes rock outcrops with woodrat sign. I designated these by markers along the highway that indicate mileage from the county line, which crosses Highway 89 close to the Sagehen road junction. Of these sites, Mile 4 (6.4 km) was the only one used for mark and recapture studies; the others are further (Miles 9.4, 9.7, 12), and were trapped occasionally as sources of live animals for the laboratory and for the introduction experiment at the Rockslide site.

At Mile 4 there are actually two rock areas separated by 0.7 km of open yellow pine forest with a few Sierra junipers. The more southern of these, Mile 3.7, is directly across the road from a heavily used Forest Service campground on the Little Truckee River. The area is a rocky slope about 300 m long, paralleling the highway, with a few scattered large boulders and a single outcrop 25 m in diameter. Wild rose grows around the outcrop, and sagebrush and bitterbrush dominate the area around the rocks. Single woodrats were caught in this area, but no breeding groups were found there.

Mile 4.2 lies north of the main campground area and consists of a 200 m long outcrop with deep crevices and extensive talus below it (plate 9a). The talus is heavily covered with bitter cherry and serviceberry; rabbit brush, sagebrush, bitterbrush, and aspen make up much of the remaining cover.

Mile 4 was the only study site where I found woodrats in all four years of my study. Breeding occurred at Mile 4.2 in three of the four years. Unfortunately, proximity to the campground caused vandalism to traps to be a continuing problem. Aside from the obvious direct effects on data collection, it also inhibited me from trapping there as often as desirable. When I did trap, I sometimes resorted to setting traps out after dark and collecting them again before dawn.

The first trapping at Mile 4 was done in mid-July 1971. At that time two adult breeding females and two subadult males were caught at Mile 4.2, but no adult male. At the same time, a single adult male (1567) was captured at two widely separated locations at Mile 3.7. Because there was little fresh sign at Mile 3.7, and 1567 was later (August and September) captured at Mile 4.2, I believe he was only temporarily at the former area and was the breeding male at Mile 4.2. In late August, 1567 was at Mile 4.2, along with the two females. The two previous subadult males were gone, but three other males and five females had appeared; no signs or rats were at Mile 3.7. Late September found the adults still there, along with two young males and two young females; an additional young male was found at Mile 3.7, apparently born at the same time (June 30) as one of the males at Mile 4.2.

In 1972 I trapped at Mile 4.2 for only two nights, at the end of August. Two breeding females were found; one had bred there the previous year also, and the other was one tagged and apparently born there the previous year. No adult male was caught, but the condition of the two females was evidence of the recent presence of one. Five young were also captured. There were fresh signs at Mile 3.7, but all four traps were destroyed the one night I attempted trapping.

In early June 1973 fresh juniper cuttings were found at Mile 4.2, but trapping produced no rats. July 1-3 a new adult male (1157) was captured alone. By late August he had been joined by a young female (1168); the two were still there at the end of September.

Female 1168 survived the winter, and in early June was trapped in breeding condition with a new adult male (1180) and three three-week old young who were likely littermates. Those three young and 1180 were missing at the end of July, but 1168 was present with four more young, born a month after the previous litter. Remaining at the end of September were the two adults, two females from the second litter, and two more females born about August 10. For the first time that summer, a large subadult male was found at Mile 3.7.

Only a small amount of fresh urine was present in mid-June of 1975, suggesting in the absence of trapping that only one woodrat was present.

OTHER FIELD DATA

Reproduction

A summary of birthdates is shown in table 15, broken down by whether the young were born in an area with one female, more than one female, or were immigrants whose home areas were unknown. The birth season runs from mid-April through mid-August. Inspection suggests that single females start and end their breeding later, but this is based on only two females. If there is a difference, it is unlikely to be a function of the area they are breeding in because both were in areas where harems had bred in other years. Attempts to correlate breeding dates with last snowfall and with snowmelt were inconclusive. All females in known breeding groups had their first litter before June 1. Snow is on the ground at Sagehen until the last week of April (weather records of Sagehen Creek Field Station), so animals must come into reproductive condition and often mate before snow is gone. Rocks may clear before surrounding areas, and in many of the study areas food plants grow within the rocky area.

The reproductive output of a female is partly a function of the number of litters she produces in a season. In the areas studied, this ranged from an average of two to three litters per female per year (table 16). There would appear to be a slight reduction in the number of litters a female produces as the number of females in the area increases, but the sample is too small to rely on.

TABLE 15

Birthdates of Young *Neotoma cinerea* Trapped at Sagehen Creek Study Sites

Parental group	April	May		June		July		August	Total
	16-30	1-15	16-31	1-15	16-30	1-15	16-30	1-15	
Single female	0	0	7[a] (41%)[b]	1 (47%)	6 (82%)	0 (82%)	1 (88%)	2 (100%)	17
Harem	2 (5%)	13 (38%)	4 (48%)	8 (68%)	6 (83%)	1 (85%)	6 (100%)	0	40
Unknown	2 (5%)	6 (21%)	10 (47%)	3 (55%)	10 (82%)	3 (89%)	4 (100%)	0	38
All	4 (4%)	19 (24%)	21 (46%)	12 (58%)	22 (82%)	4 (86%)	11 (98%)	2 (100%)	95

a. Number of young born in interval

b. Cumulative percentage

TABLE 16
Litters of Neotoma cinerea
Born at Sagehen Creek Study Sites

Area	Year	Number of females	Number of litters	Litters/Female
Spring	1971	1	3	3.0
Mile 4	1974	1	3	3.0
Mile 4	1971	2	6	3.0
Mile 4	1972	2	(4)[a]	-
Spring	1972	2	5	2.5
X89	1973	3	6	2.0

a. Data incomplete

Dispersal

Bushy-tails may disperse at any age from shortly past weaning to several years old, as described above. In the case of young animals, their second month is the period which, in the laboratory, is marked by increasing strife with other young and with parents. In the field, as noted repeatedly, young may move out to a peripheral den before disappearing entirely. A similar observation has been made for the young of N. fuscipes (Cranford, 1977).

In four cases a marked animal appeared elsewhere (arrows on fig. 19). The longest recorded disperser was a female who moved 3.2 km in less than 28 days (see X89 account above). To my knowledge, this is the longest distance reported for a woodrat of any species. Additionally, a young male moved 2.2. km in 10 days, also surpassing previous records. Smith (1965) reported a movement of 1.6 km for N. fuscipes.

Of 40 immigrant subadults captured, only two (1120 and 1150) could be identified as to their birthplace. While some of the rest may have originally come from home areas I studied, but were not tagged, certainly not all did. My own thorough searches of the basin and surrounding area, as well as the extensive trapping efforts of others, suggest that many of the immigrants could not have been born in the basin, and must have moved even greater distances than the ones measured.

Rock areas can be roughly divided into two categories of "quality" for woodrats. A "breeding" area is one in which survival over the

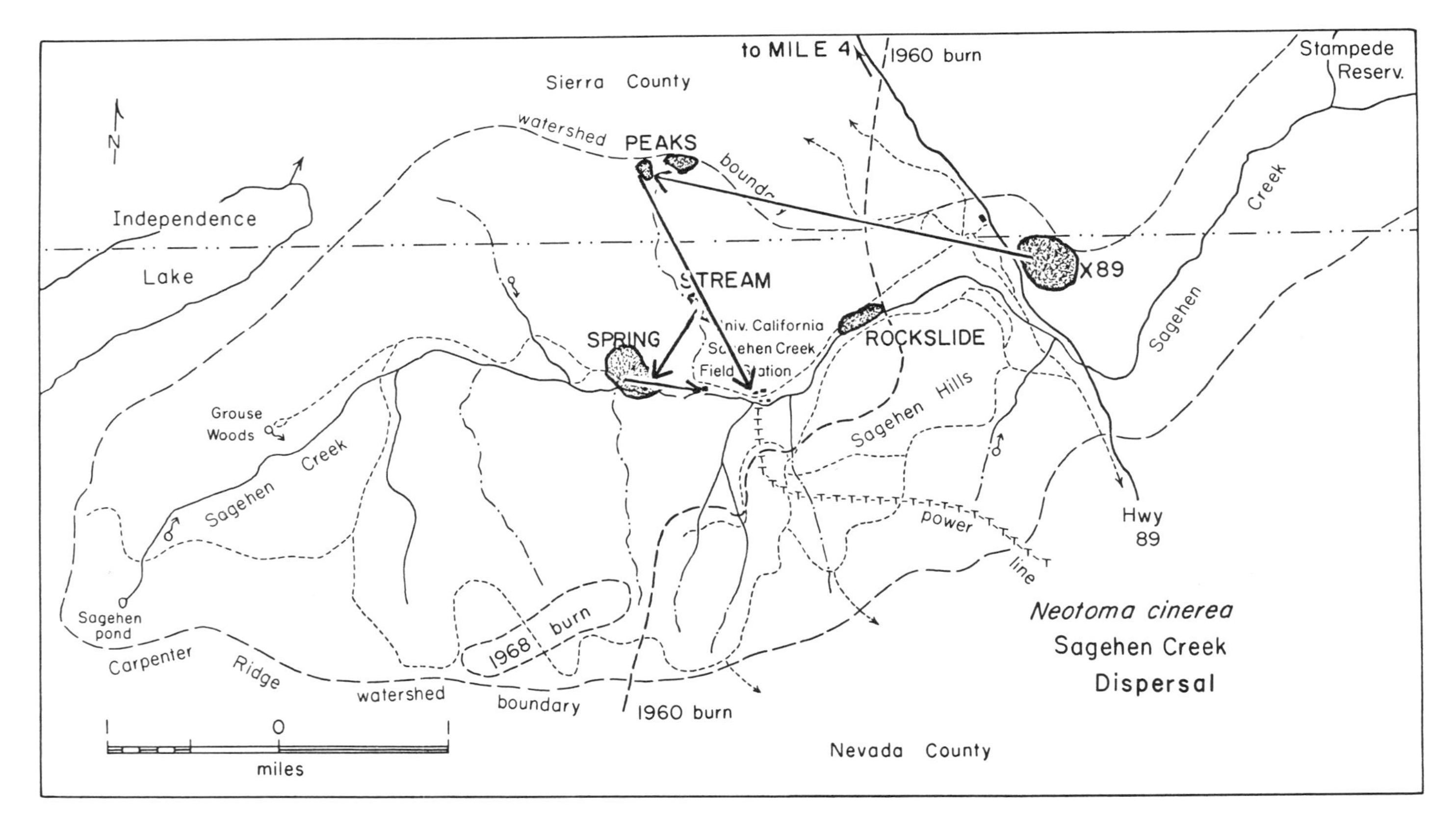

Figure 19. Straight-line routes (solid arrows) of known dispersing _Neotoma cinerea_ in the Sagehen Creek basin.

winter is possible for several woodrats and thus the residents will survive to breed. Such an area is large in size, and has numerous good sites for dens under large boulders or in deep crevices. Spring and X89 sites fit this category. Although no breeding took place at the Rockslide during the study, I include it as a breeding area based on the amount of rock cover it can provide and on the extent of old urine marking. The other category, "nonbreeding," is given to rock areas too small and poor in shelter to provide for many woodrats, such as Peaks or Stream sites.

A young woodrat has four choices: it can remain in the home area, find a vacant breeding area, find a nonbreeding area, or move into an occupied breeding area. The first choice is almost entirely limited to females. If we compare the remaining choices with the animals taking them, an interesting relationship appears (fig. 20). Immigrants settling in vacant breeding areas are almost exclusively young born before June 1, and thus members of the first litters of the season. Animals settling in the nonbreeding areas are almost all born after June 1. (The relationship between birthdates of settlers in these two types of areas is significant at $p<.001$; Chi-square = 12.6). Immigrants into occupied breeding areas are few and late-born, but they apparently survive well if they can settle there at all. Settlers in nonbreeding areas either do not remain there or perish over the winter, while those in vacant breeding areas have a better chance. This pattern is reinforced by the exclusion of later immigrants by prior occupants of the same sex, as at the Spring in 1973, and at X89 in 1974.

On a number of occasions, several immigrants appeared at the same time and had estimated birthdates so close together that they could easily have been littermates. (Litter dates for different females are not synchronized, even within the same rock area.) Since young woodrats play together a lot, and are known to move around their home areas together, it is reasonable to speculate that groups of siblings may disperse together. On settling in an area, patterns typical of adult harems appear, with separate male territories or only a single male, and overlapping female home ranges.

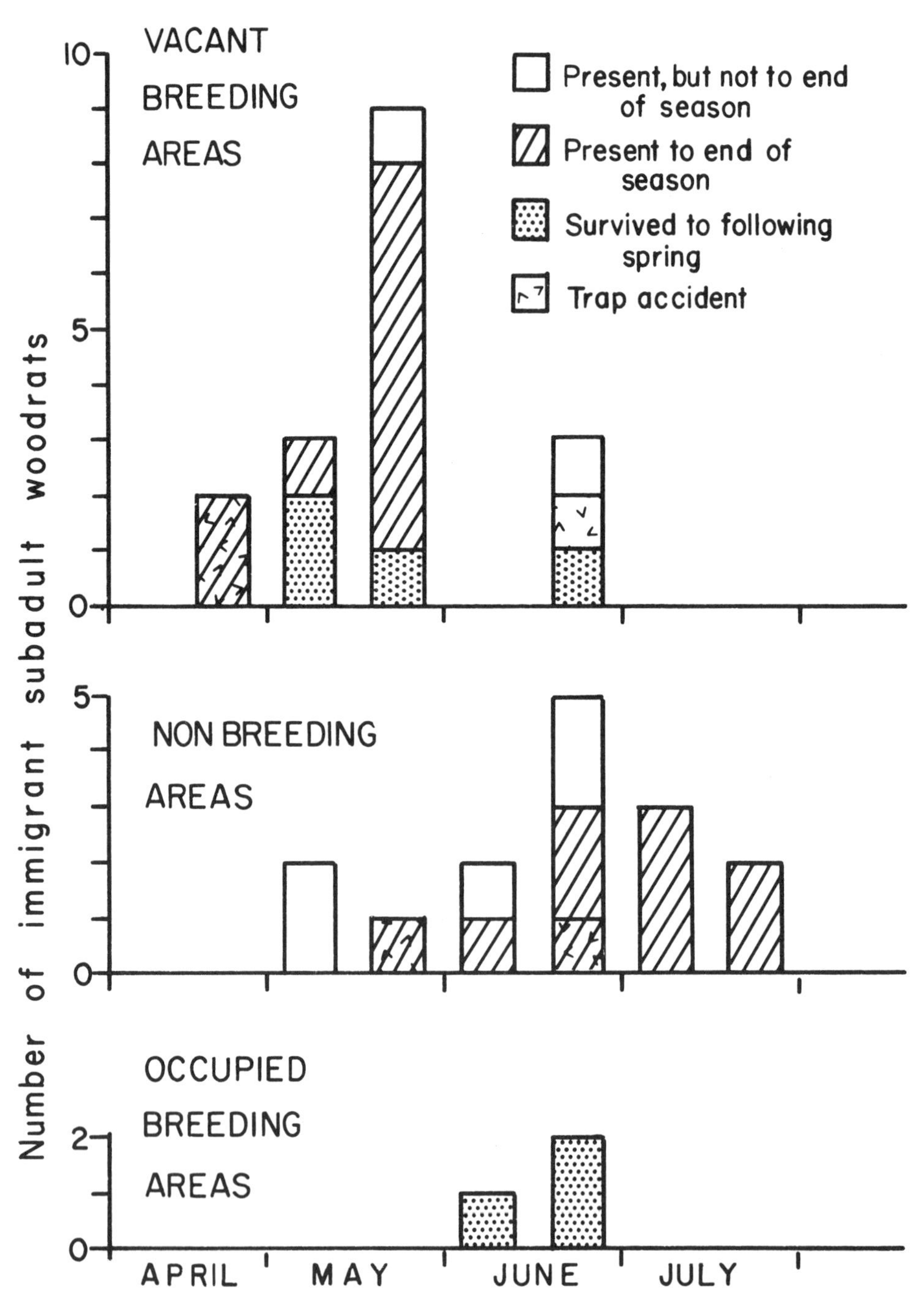

Figure 20. Birthdates of first-year Neotoma cinerea settling in different kinds of rock areas.

Food

I did not study food habits in detail, for Finley's (1958) study more than amply demonstrated the species' flexibility in diet. To study food adequately, I would have had to break open dens, which would have been antithetical to studying an undisturbed population.

So long as there is reasonable plant growth of almost any kind, bushy-tails would seem unlikely to be limited by food. The observations I made confirmed this flexibility. Extensive food storage occurred only in late August and September, before the leaves turned color. Leaves were collected and placed in crevices where the sun dried them quickly; as a consequence of rapid drying, the leaves stay green, and can be found that way in midwinter. The leaves collected varied depending on availability. At the Lower Spring one year I observed storage of aspen and wild onion. In another year, aspen, dogwood, alder, and alderleaf coffeeberry were collected there. At X89, bitter cherry leaves were stored. Aspen was the major store at Mile 4.2 and at the Upper Spring.

Survival

N. cinerea does not breed until the spring and summer following birth. An important statistic, therefore, is the survivorship after weaning. An index of survival rates after this point is the proportion of adults in a population, assuming reproductive rates are similar in populations being compared and that mortality rates are similar within populations for subadults and adults.

Examining the field population first, we find that of 116 animal-summers, 22 were adults, yielding a proportion of 19% adults. I am making the simplifying assumption of treating each summer as a new sample independent from preceding and following summers. The amount of population movement (see dispersal, above) makes such an assumption reasonable. This assumption also makes the statistic comparable to the museum samples, which did not distinguish between age classes of adults.

Data concerning relative reproductive rates and adult survival are few. As described in the museum section, different age classes of adults are probably not reliably separable, thus yielding no data on adult survival. The sample of adult survival from the field population is too small for statistical testing, but at least suggests that adult survival rates may be similar to that for young animals: the 22 adult animal-summers above represent 18 individuals, three of which survived a second year (17%); of those three, one (33%) survived a third year. Lacking better data, I assume that similar mortality factors are operating at similar rates between years of adulthood as are operating prior to it, and that such may be the case for the various subspecies. Bushy-tails are physiologically capable of living longer, for at least 14 of 75 lab animals survived three years or more, although those older animals might be less fit for survival in the wild.

Comparative reproductive rates are equally problematical. Egoscue (1962) published laboratory breeding data for *N. c. acraia*, and I include data (table 9) for *N. c. alticola*. Egoscue apparently observed a higher reproductive rate in his colony, but I suspect that much of that was due to better nutrition he was able to supply, as well as other conditions, for breeding success declined in my colony each year. Once this factor is taken into account, I doubt that differences in reproductive rates are sufficiently great between these subspecies to account for the differences we see below in the proportion of adults in the population. Data for *N. c. pulla* are almost nonexistent. Factors in addition to milder weather may operate on this subspecies, for its range includes less rock habitat, and it must depend on other shelter. J. H. Brown (1968) described the similar habitat of *N. c. fusca*, from the Oregon coast.

Turning now to the museum populations, we find that the proportion of adults in the field population is only half that in the museum sample of *N. c. alticola* (table 2c), the same subspecies, in which a total of 463 specimens included 178 adults, giving a proportion of 38.4% (Chi-square, corrected for continuity = 14.7; $p<.001$). This difference is not surprising, for two reasons. First, by concentrating my trapping for *N. cinerea* at the end of the breeding season, I probably biased the field sample to first-year animals, whereas the museum collections are more evenly spaced from spring to fall. Secondly, the field population was subject to added mortality from accidents during live-trapping (fig. 20), and, on two occasions (three animals), to inadvertent collecting by other workers. Thus, the survival rate seen in the museum population may be more representative for *N. c. alticola*. (Differential collection or retention of adults by museum collectors could have produced bias in those samples, but that factor, if real, should have operated similarly for the three subspecies.)

Both of the other subspecies analyzed show significantly higher proportions of adults. N. c. acraia (table 2b) had 48.7% adults (130 adults in 267 specimens), and N. c. pulla (table 2d) had 53.2% adults (74 adults in 139 specimens). Both of these differ significantly ($p<.01$) from the museum sample of N. c. alticola (Chi-square, corrected for continuity = 6.87 and 9.01, respectively). They do not differ from each other (Chi-square, corrected for continuity = 0.58).

These data suggest that the survival rate of N. c. alticola is 10-15% poorer than that of the other two subspecies. The causes of mortality are often difficult to identify in a field population, but most fall into the categories of predation, disease, food shortage, or adverse weather.

Of these, weather is the one most obviously different for the three subspecies. N. c. alticola, occurring both inland and further north (fig. 2), is faced with considerably more severe winters than are the other two subspecies. J. H. Brown (1968) measured winter temperatures from dens of montane, high desert, and coastal populations of N. cinerea as well as from houses of N. albigula, the white-throated woodrat. These demonstrated the value of rock dens as protection from extreme winter cold. Although he was working with populations different from those of the present study, the general principle is applicable: in the coldest climate, bushy-tails must have adequate rock shelter to survive through the winter. With one exception (1598 at the Stream site), woodrats were not known to have survived the winter at Sagehen unless they were in a "good" rock area the previous fall (see also dispersal section). Even in the best areas, losses over the winter were high, and disappearance of all or most members of an active settlement between October and June was common. The scarcity of adequate rock shelter would argue for this as the most important limiting factor for the bushy-tail population.

Predation, of course, takes its toll. Weasels (Mustela frenata) are seen regularly throughout the Sagehen basin, and I cited above (X89 site) a case where I radiotracked an apparent weasel victim. Martens (Martes americana) are seen in the basin, especially in winter, and are known to take bushy-tails occasionally (Murie, 1961). Bailey (1898) found bushy-tail remains in the scat of bobcats (Lynx rufus), a species which is not uncommon at Sagehen (Hawthorne, 1970), although rarely seen. Hawthorne found no evidence of woodrat predation in the scats of coyotes (Canis latrans) at Sagehen, but the low abundance of bushy-tails relative to other potential prey species would make such predation difficult to detect in this way. Potential avian predators include the great horned owl (Bubo virginianus), which I heard often near the study sites, and the diurnal red-tailed hawk (Buteo jamaicensis) and goshawk (Accipiter gentilis). Although losses of woodrats to these predators may be significant, and another reason for their requiring good shelter, there is little to suggest that predation could be responsible for the difference in survivorship between the subspecies.

Food, as described in the previous section, is a far more abundant resource than is shelter. Only two circumstances suggest themselves as possibly food limiting. In an abnormally long winter, stored food supplies might run out, although bushy-tails do include in their diet buds, young shoots, green bark, and leaves of juniper and several nondeciduous shrubs (present study; Sumner and Dixon, 1953; Finley, 1958), all of which could be available in winter. If stored food were to become limiting in this case, an important factor would be the amount of shelter available and suitable for food storage, probably more so than the amount of food initially available for storage; thus we are again faced with shelter as the limiting factor. The other case for food limitation might be made for the rare circumstances where extensive rock shelter is available. This case is sufficiently unusual to be important only locally and is discussed in the following section on continuous habitat.

The role of disease in woodrat mortality is difficult to assess, although it may be one factor promoting dispersal and formation of new settlements. Plague can decimate local populations of woodrats periodically (Nelson and Smith, 1976). The disease was first discovered in the United States about 1900 in San Francisco; by 1910, it was well established in wild rodent populations of the California coastal region, and over the next 25 years spread through the western states (Eskey and Haas, 1940; Link, 1950; Pollitzer, 1954). Murray (1963) pointed out that plague tends to persist enzootically in "relatively cold upland sites," which accurately describes the range of N. cinerea.

On the other hand, continuing surveys of plague suggest that it exists in relatively small pockets within the total range (Eskey and Haas, 1940; Link, 1950; Murray, 1963). Also, if plague were causing higher mortality, we should see a difference in the proportion of adults in the samples from the two museums. Most of the USNM specimens of N. c. alticola were collected prior to the known spread of plague, and contain 40% adults (107 adults/265 total). The MVZ materials were collected after, or at least concurrent with that spread, and contain 36% adults (71 adults/198 total). These proportions are not significantly different (Chi-square, corrected for continuity = 0.796), and thus introduction of plague seems not to have affected the survival rate. Similarly, we would not expect plague to be responsible for differences in survival between subspecies.

Continuous Habitat

At Sagehen, the suitable habitat of rock outcrops occurs in isolated islands so that each woodrat settlement is well separated from the next. What happens when rock is more extensive, so that more shelter is available?

Several sources had suggested that such habitat was abundant in northeastern California (Modoc County), so I spent 10 days there in July and August 1974 searching for sites to study such populations.

Although geologically recent volcanic activity has contributed large areas of exposed rock there, less of it was suitable than I had hoped. With the exception of deep lava tubes (cf. Nelson and Smith, 1976), the large flat lava-flow areas are not sufficiently broken up to provide much shelter.

Lava rims provided extensive shelter in a few places. In some, large talus slopes (plate 9b) appeared suitable, but contained few urine marks, fresh or old. Trapping produced only small numbers of bushy-tails, confirming the low population estimates. Possibly in a rock area so large, food does become limiting, because the only food plants available grow around the edge. Consequently, the population might reach only low densities.

Elsewhere, narrower lava rims extended along edges of plateaus, creating a relatively continuous linear denning habitat. Examining these more closely, I found considerable variability in the degree of brokenness, and "good" habitat alternated with relatively poor. Even in better habitats here there were few rats, and many areas were abandoned. Recurrent plague epizootics (Nelson and Smith, 1976) may have contributed to the low numbers.

Thus, even in apparently continuous denning habitat, contiguous settlements are probably uncommon. Judging from the occasional presence at Sagehen of additional males near but not in breeding settlements, separate harems may exist near each other without losing their identities where suitable continuous habitat exists.

As discussed above, in the section on survival, the rarity of such continuous habitat would make it a minor factor in the usual social organization of the species. I would expect the more common case to be like Sagehen, where the scarcity of adequate rock shelter would make that the primary limiting resource.

DISCUSSION

The social system of the bushy-tailed woodrat demonstrates the interaction between evolutionary background and more immediate environmental factors discussed by Crook et al. (1976). Several of the principles used by Geist (1974) in relating social evolution to ecology of ungulates prove applicable to the neotomine-peromyscine rodents in general and to *N. cinerea* in particular.

Geist, citing Jarman (1968) and Bell (1971; see also Bell, 1970), started with the point that body size is related to diet of a species. Smaller species feed on higher quality (more digestible) forage; they need less "digestive apparatus" and less food bulk, and therefore can be smaller. On the other hand, the scarcity of such high-quality food enforces the smaller size, as well as preventing high biomass for the population, and requires them to search widely for the dispersed food supply. Small size also makes greater metabolic demands due to surface/volume relationships, further mandating a high-quality diet. Conversely, larger ungulates are those with an abundant but less digestible diet. In both cases the relationship is a close one, for in each the diet not only permits but requires the size, and size does the same for diet.

Although Geist referred this argument specifically to ungulates, it can be extended to neotomine-peromyscine rodents. Meserve's (1976) data for diets of four coexisting cricetine rodents in a California sage community fit the model. *Peromyscus californicus*, the largest, ate the most vegetation; *Reithrodontomys megalotis*, the smallest, had the most animal food (insect larvae and other arthropods). Two other *Peromyscus* species were intermediate both in size and diet. *Neotoma*, with a diet largely of leaves (Finley, 1958; Meserve, 1974), is still larger, as the model would predict. Its large cecum (Howell, 1926) and its habit of reingesting feces (present study) also attest to its low quality diet.

Geist continued by relating body size of ungulates to strategy for predator defense. The larger rodents are obviously not comparably dangerous to their predators as are large ungulates. For a rodent, large size may be a liability relative to predators. Not only does it represent a large meal for a given hunting effort, but its large size makes it more obvious and less able to be secretive. The evolutionary response in this relatively arboreal group, unlike that of ungulates, has been to build large houses. The ready return of a woodrat to its house when threatened,and its unwillingness to leave it, both indicate how important the house is against predators. These houses not only provide predator protection but also protect from extremes of heat and cold (Lee, 1963; J. H. Brown, 1968; Brown and Lee, 1969). Not only does house-building characterize *Neotoma*, but *Peromyscus californicus* also uses relatively large and complex houses (McCabe and Blanchard, 1950). The abundant low quality diet (leaves), which led to the large house and predator problems, also allow an individual to remain in one place and to expend considerable energy in building a house.

The house is a defensible resource. Energy is required in building, maintaining, and defending it. Possession of a house is so important to woodrat survival that a high level of intraspecific aggression and solitary house occupancy are basic to the genus.

From this evolutionary heritage, *N. cinerea* became the boreal representative. The morphological differences, mostly those adapting it to colder climates (insulation and changes related to larger size), are sufficient to place it in a separate subgenus (Goldman, 1910).

The museum data are consistent with the hypothesis that climate is important, even within the species. The smallest subspecies, *N. c. pulla*, occurs in the coastal region and at lower elevations, where one would expect the mildest climate (cf. J. H. Brown, 1968; Brown and Lee, 1969). Of the two inland subspecies studied, *N. c. acraia* occurs further south and is the next largest. The largest subspecies of those examined, *N. c. alticola*, is the more northern inland subspecies, occurring where winters are most severe. It also has a survivorship to adulthood up to 15% less than that of the two milder-weather subspecies, demonstrating the severity of the selection still operating on it.

The most important resource for Neotoma cinerea is adequate rock shelter for winter surival. Eighty percent of the woodrat specimens in the museums were collected with other woodrats, demonstrating the degree to which they aggregate, in spite of the aggressive nature they share with other species of their genus.

The typical woodrat house of sticks (plate 1b) apparently is inadequate in an area of extreme cold and heavy snow. (N. cinerea may use stick houses in low-elevation areas where there is little snowfall (Grinnell et al., 1930).) Not only do the rocks provide better insulation (see survival section, above), but they may be more impermeable to blowing snow. Although typical woodrat houses are relatively waterproof, drifting snow might be better able to penetrate them. N. floridana magister, the most northern subspecies of the eastern woodrat, is almost exclusively a rock dweller also (Newcombe, 1930; Poole, 1940), although elsewhere the species typically uses a stick house (Lay and Baker, 1938; Rainey, 1956). Fitch and Rainey (1956) attributed the decimation of a population of N. floridana attwateri in Kansas not only to two extremely cold winters but also to frozen sleet which penetrated stick houses. As discussed above (in survival section), over most of the range of N. cinerea the lack of adequate rock shelter is probably more limiting than is the amount of food available.

It follows from the trapping summaries that extinction of local populations is a common event. The clumping of many individuals of a large rodent may prove inviting to predators such as weasels (present study) and martens (Murie, 1961) which can follow the woodrats into the rocks. For an individual woodrat's genes to survive these frequent local extinctions, whatever the cause, it is advantageous to send out dispersing offspring as well as allowing some to remain to take over the home area (cf. Van Valen, 1971).

The question, however, is why bushy-tails should live in harems. The advantage to the male of excluding other males is obvious. If he can father all litters born, more of his genes are passed on. The male has little to do with raising the young, so his energetic contribution to additional litters is low, and extra females gain him far more than they cost him, so long as his defense efforts remain successful and do not lead to his demise or expulsion.

Females, on the other hand, share breeding areas because it may cost them more to exclude other females than they gain. The pressure for them to be in a breeding area (presumably meaning there are enough den sites to go around) is extreme, for if they are not in a sufficiently good area, they will not survive the winter to breed. Sharing an area with other females may require some sacrifice (cf. table 16), but all breed and produce young. Orians (1969), primarily in reference to birds, also made the point that restricted nest sites and a widespread food supply are conducive to polygyny. At the time of weaning and dispersal, weather constraints are lifted, so survival is possible for the young in less sheltered areas. If there is a reduction in the number of litters, it is in the later litters, for all females have given birth by June 1. Those early litters have been shown to be the most successful dispersers, and the most likely offspring to carry on a female's genes. Late young are less likely to survive, so if a female fails to produce a later litter, as a result of sharing a breeding area, it is a minor sacrifice compared to the risk of failing to breed as a result of excessive attempts to exclude other females.

Another factor which may mitigate possible disadvantages of sharing is that females in a harem are often related. I discussed several cases of harems consisting of sibling females or of mother and daughter. Also, if siblings disperse together, as is suggested by trapping data for immigrants, new settlements may contain sisters born elsewhere. In either case, by allowing closely related females to produce young, the female's inclusive fitness (W. D. Hamilton, 1964; Maynard Smith, 1964; Wilson, 1975) would be increased and select for tolerant behavior.

Food stored for the winter is another resource that figuratively may be "shared" and may be considered a cooperative venture in the sense that if several animals collect the food, but not all survive, the others can benefit by using the extra food. If those benefiting from it are close relatives, inclusive fitness of the collector would also be increased. Whether related or not, the possibility of benefiting from another's food stores later in the winter may be another factor in inducing bushy-tails to live near each other.

Finally, I started with the question of why scent marking should be enhanced in N. cinerea relative to other woodrats. Urine probably functions as a trail marker for a species which spends much of its time in completely dark habitat. Although a woodrat relies on kinesthetic memory when in a hurry, its memorization of that path is probably enhanced by following a trail of urine odor at lower speed. Kinsey (1976) noted that N. floridana magister leaves urine trails. He found that individuals released into a clean new cage are disoriented compared to individuals newly released into a previously marked cage. In addition, urine may also perform the function of identifying the sex, age, and other characteristics of its depositer, and urine posts may then act as information exchanges. Nelson and Smith's (1976) suggestion that the ventral gland is used primarily in trail marking has the disadvantage that in that case it should not be sexually dimorphic.

Ralls (1971) pointed out that scent marking can be a form of display in a mammal, and that would seem to be the case for the ventral gland of N. cinerea. In polygamous animals of many kinds, the instrument of display is enhanced by sexual selection both to aid in threatening or competing with others of the same sex and to attract the opposite sex (Darwin, 1871). Whether we are discussing plumage and calls of birds, the large size of pinnipeds, or the scent gland of a woodrat, the principle is the same.

The trend to greater size dimorphism in the subspecies of N. cinerea that is largest and lives in the most severe climate provides further evidence for the environmental origin of its polygyny. Although large size aids in survival in cold climates, physiology alone should not dictate sexual dimorphism in this feature. A polygynous social system which resulted from the severity of the climate would cause, by male competition, the observed cline of dimorphism. Secondarily, physiology might be important because the extra males, which are excluded from the best areas, might survive better if larger.

Literature Cited

ARATA. A. A.

1964 The anatomy and taxonomic significance of the male accessory reproductive glands of muroid rodents. Bull. Florida State Mus., 9:1-42.

BAILEY, V.

1898 Unpublished field notes from Arc Dome, Nevada, May 24-26. At U.S. National Museum, Washington, D.C.

1918 The mammals. In V. Bailey and F. M. Bailey (eds.), Wild animals of Glacier National Park. U.S. Dept. Interior, Natl. Park Service, pp. 25-102.

1930 Animal life of Yellowstone National Park. Springfield, Ill., and Baltimore: Charles C. Thomas.

1936 The mammals and life zones of Oregon. N. Amer. Fauna, 55:1-416.

BALGOOYEN, T. G.

1976 Behavior and ecology of the American Kestrel (<u>Falco sparverius</u> L.) in the Sierra Nevada of California. Univ. Calif. Publ. Zool., 103:1-87.

BANFIELD, A. W. F.

1974 The mammals of Canada. Toronto: Univ. Toronto Press.

BEAVER, D. L.

1972 Avian species diversity and habitat use in forests of the Sierra Nevada, California. Ph.D. diss., Univ. California, Berkeley.

BELL, R. H. V.

1970 The use of the herb layer by grazing ungulates in the Serengeti. In: Animal populations in relation to their food resources, A. Watson, ed. (1969 symposium of Brit. Ecol. Soc.). Oxford: Blackwell Sci. Publ., pp. 111-124.

1971 A grazing ecosystem in the Serengeti. Sci. Amer., 225(1):86-93.

BIRNEY, E. C.

1973 Systematics of woodrats (genus _Neotoma_) in central North America. Univ. Kansas Mus. Nat. Hist., Misc. Publ., 58:1-173.

BOCK, J. H., and C. E. BOCK

1969 Natural reforestation in the northern Sierra Nevada--Donner Ridge burn. Proc. Ann. Tall Timbers Fire Ecol. Conf., 9:119-126.

BOLEY, R. B., and T. E. KENNERLY, Jr.

1969 Cellulolytic bacteria and reingestion in the plains pocket gopher, _Geomys bursarius_. J. Mamm., 50:348-349.

BOX, T. W.

1959 Density of plains wood rat dens on four plant communities in south Texas. Ecology, 40:715-716.

BRONSON, F. H., and D. CAROOM

1971 Preputial gland of the male mouse: attractant function. J. Reprod. Fert., 25:279-282.

BROWN, A. M.

1976 Ultrasound and communication in rodents. Comp. Biochem. Physiol. A, 53:313-317.

BROWN, J. C., and J. D. WILLIAMS

1972 The rodent preputial gland. Mammal Rev., 2:105-147.

BROWN, J. H.

1968 Adaptation to environmental temperature in two species of woodrats, _Neotoma cinerea_ and _N. albigula_. Univ. Mich. Mus. Zool., Misc. Publ., 135:1-48.

BROWN, J. H., and A. K. LEE

1969 Bergmann's rule and climatic adaptation in woodrats (_Neotoma_). Evolution, 23:329-338.

BROWN, J. H., G. A. LIEBERMAN, and W. F. DENGLER
1972 Woodrats and cholla: dependence of a small mammal population on the density of cacti. Ecology, 53:310-313.

BURNETT, J. L., and C. W. JENNINGS
1962 Geologic map of California, Chico sheet, 1:250,000. Calif. Dept. Conservation, Div. Mines and Geology.

BYRNE, S. B.
1972 Dominance relationships and comparative ecology of two sympatric chipmunks (Eutamias). M.A. thesis, Univ. California, Berkeley.

CAMERON, G. N.
1971 Niche overlap and competition in woodrats. J. Mamm., 52:288-296.

CAMPBELL, B. G. (ed.)
1972 Sexual selection and the descent of man, 1871-1971. Chicago: Aldine.

CHAPMAN, A. O.
1951 The estrous cycle in the woodrat, Neotoma floridana. Univ. Kansas Sci. Bull., 34:267-299.

COCHRAN, W. W.
1966 Some notes on the design of a directional loop antenna for radiotracking wildlife. BIAC Information Module no. M4. Washington, D.C.: Amer. Inst. Biol. Sciences, Biointrumentation Advisory Council.

COCHRAN, W. W., and R. D. LORD, Jr.
1963 A radio-tracking system for wild animals. J. Wildl. Mgmt., 27:9-24.

CRANFORD, J. A.
1970 Home range of the dusky-footed woodrat, Neotoma fuscipes. M.A. thesis, San Francisco State Univ., California.
1977 Home range and habitat utilization by Neotoma fuscipes as determined by radiotelemetry. J. Mamm., 58:165-172.

CROOK, J. H., J. E. ELLIS, and J. D. GOSS-CUSTARD
1976 Mammalian social systems: structure and function. Anim. Behav., 24:261-274.

DALQUEST, W. W.

1948 Mammals of Washington. Univ. Kansas Publ., Mus. Nat. Hist., 2:1-444.

DARWIN, C. R.

1871 The descent of man and selection in relation to sex. London: John Murray.

DAVIS, J. A., Jr.

1970 Underworld character: the Allegheny packrat. Anim. Kingdom, 73(2):25-29.

DAVIS, W. B.

1939 The Recent mammals of Idaho. Caldwell, Idaho: Caxton.

DEWSBURY, D. A.

1972 Patterns of copulatory behavior in male animals. Q. Rev. Biol., 47:1-33.

1974a Copulatory behavior of white-throated wood rats (Neotoma albigula) and golden mice (Ochrotomys nuttalli). Anim. Behav., 22:601-610.

1974b Copulatory behavior of Neotoma floridana. J. Mamm., 55:864-866.

1975 Diversity and adaptation in rodent copulatory behavior. Science, 190:947-954.

DIXON, J.

1919 Notes on the natural history of the bushy-tailed wood rats of California. Univ. Calif. Publ. Zool., 21:49-74.

EGOSCUE, H. J.

1957 The desert woodrat: a laboratory colony. J. Mamm., 38:472-481.

1962 The bushy-tailed wood rat: a laboratory colony. J. Mamm., 43:328-337.

EISENBERG, J. F.

1959 A study of intra-specific social behavior among mice of the genus Peromyscus. M.A. thesis, Univ. California, Berkeley.

1963 The behavior of heteromyid rodents. Univ. Calif. Publ. Zool., 69:1-100.

1966 The social organizations of mammals. Handb. der Zool., VIII, 10(7):1-92.

EISENBERG, J. F., and D. G. KLEIMAN

1972 Olfactory communication in mammals. Ann. Rev. Ecol. Syst., 3:1-32.

EMLEN, S. T., and L. W. ORING

1977 Ecology, sexual selection, and the evolution of mating systems. Science, 197:215-223.

ESKEY, C. R., and V. H. HAAS

1940 Plague in the western part of the United States. Publ. Health Bull., 254:1-83.

ESTEP, D. Q., and D. A. DEWSBURY

1976 Copulatory behavior of Neotoma lepida and Baiomys taylori: relationships between penile morphology and behavior. J. Mamm., 57:570-573.

EWER, R. F.

1968 Ethology of mammals. London: Logos Press.

FINLEY, R. B., Jr.

1958 The wood rats of Colorado: distribution and ecology. Univ. Kansas Publ., Mus. Nat. Hist., 10:213-552.

FITCH, H. S., and D. G. RAINEY

1956 Ecological observations on the woodrat, Neotoma floridana. Univ. Kansas Publ., Mus. Nat. Hist., 8:499-533.

GANDER, F. F.

1929 Experiences with wood rats, Neotoma fuscipes macrotis. J. Mamm., 10:52-58.

GEIST, V.

1974 On the relationship of social evolution and ecology in ungulates. Amer. Zool., 14:205-220.

GOLDMAN, E. A.

1910 Revision of the wood rats of the genus Neotoma. N. Amer. Fauna, 31:1-124.

GRAY, J.

1953 How animals move. Cambridge, Engl.: Cambridge Univ. Press.

GRINNELL, J., J. DIXON, and J. M. LINSDALE

1930 Vertebrate natural history of a section of northern California through the Lassen Peak region. Univ. Calif. Publ. Zool., 35:1-594.

GRINNELL, J., and T. I. STORER
1924 Animal life in the Yosemite. Berkeley: Univ. Calif. Press.
HALL, E. R., and K. R. KELSON
1959 The mammals of North America (2 vol.). New York: Ronald Press.
HAMILTON, W. D.
1964 The genetical theory of social behaviour, I., II. J. Theoret. Biol., 7:1-52.
1971 Geometry for the selfish herd. J. Theor. Biol., 31:295-311.
HAMILTON, W. J., Jr.
1953 Reproduction and young of the Florida wood rat, Neotoma f. floridana (Ord). J. Mamm., 34:180-189.
HAWTHORNE, V. M.
1970 Movements and food habits of coyotes in the Sagehen Creek basin and vicinity. M.S. thesis, Univ. Nevada, Reno.
HOOPER, E. T.
1940 Geographic variation in bushy-tailed wood rats. Univ. Calif. Publ. Zool., 42:407-424.
1960 The glans penis in Neotoma (Rodentia) and allied genera. Univ. Michigan, Mus. Zool., Occas. Paper, 618:1-21.
1968 Classification. In: Biology of Peromyscus (Rodentia), J. A. King, ed., Amer. Soc. of Mammalogists, Special Publ. no. 2, pp. 27-74.
HORVATH, O.
1966 Observation of parturition and maternal care of the bushy-tailed wood rat (Neotoma cinerea occidentalis Baird). Murrelet, 47:6-8.
HOWELL, A. B.
1926 Anatomy of the wood rat. Baltimore: Williams and Wilkins.
JARMAN, P.
1968 The effect of the creation of Lake Kariba upon the terrestrial ecology of the middle Zambezi Valley, with particular references to the large mammals. Ph.D. diss., Manchester Univ., Engl. (Not seen; cited by Geist, 1974.)

KALUGIN, Yu. A.

1974 (Coprophagy of rodents and double-toothed rodents and its physiological importance.) Zool. Zh., 53:1840-1847. (In Russian, with English summary.)

KING, J. A.

1955 Social behavior, social organization, and population dynamics in a blacktailed prairie dog town in the Black Hills of South Dakota. Univ. Michigan, Contrib. Lab. Vert. Biol., 67:1-123.

KINSEY, K. P.

1971 Social organization in a laboratory colony of wood rats, *Neotoma fuscipes*. In: Behavior and environment: the use of space by animals and men, A. H. Esser, ed. New York: Plenum Press, pp. 40-45.

1972 Social organization in confined populations of the Allegheny woodrat, *Neotoma floridana magister*. Ph.D. diss., Bowling Green State Univ., Ohio.

1976 Social behaviour in confined populations of the Allegheny woodrat, *Neotoma floridana magister*. Anim. Behav., 24:181-187.

KLEIMAN, D. G.

1971 The courtship and copulatory behavior of the green acouchi, *Myoprocta pratti*. Z. Tierpsychol., 29:259-278.

KNOCH, H. W.

1968 The eastern wood rat, Neotoma floridana osagensis: a laboratory colony. Trans. Kansas Acad. Sci., 71:361-372.

LAWRENCE, B.

1941 Incisor tips of young rodents. Field Mus. Nat. Hist., Zool. Ser., 27:313-317.

LAY, D. W. and R. H. BAKER

1938 Notes on the home range and ecology of the Attwater wood rat. J. Mamm., 19:418-423.

LEE, A. K.

1963 The adaptations to arid environments in wood rats of the genus *Neotoma*. Univ. Calif. Publ. Zool., 64:57-96.

LEYHAUSEN, P.

1965 The communal organization of solitary mammals. Symp. Zool. Soc. London, 14:249-263.

LINK, V. B.

1950 Plague. CDC Bull., 9(8):1-7.

LINSDALE, J. M., and L. P. TEVIS, Jr.

1951 The dusky-footed wood rat. Berkeley and Los Angeles: Univ. Calif. Press.

MACKAY, R. S.

1970 Bio-medical telemetry, 2nd ed. New York: John Wiley & Sons.

MARTELL, A. M., and J. N. JASPER

1974 A northern range extension for the bushy-tailed wood rat, Neotoma cinerea (Ord). Canad. Field Nat., 88:348.

MARTIN, R. J.

1973 Growth curves for bushy-tailed woodrats based upon animals raised in the wild. J. Mamm., 54:517-518.

MAYNARD SMITH, J.

1964 Group selection and kin selection. Nature (London), 201:1145-1147.

McCABE, T. T., and B. D. BLANCHARD

1950 Three species of Peromyscus. Santa Barbara, Calif.: Rood Assoc.

MESERVE, P. L.

1974 Ecological relationships of two sympatric woodrats in a California coastal sage scrub community. J. Mamm., 55:442-447.

1976 Food relationships of a rodent fauna in a California coastal sage scrub community. J. Mamm., 57:300-319.

MOIR, R. J.

1968 Ruminant digestion and evolution. In: Handbook of physiology, E. F. Cade, ed. Washington, D.C.: Amer Physiol. Soc., vol. 5, sec. 6, pp. 2673-2694.

MUGFORD, R. A., and N. W. NOWELL

1971 The preputial glands as a source of aggression-promoting odors in mice. Physiol. Behav., 6:247-249.

MUIR, J.

1901 Our national parks. Boston: Houghton Mifflin.

MURIE, A.

1961 Some food habits of the marten. J. Mamm., 42:516-521.

MURRAY, K. F.

1963 An ecological approach to a plague program in California. Calif. Vector Views, 10:13-17.

MURRAY, K. F., and A. M. BARNES

1969 Distribution and habitat of the woodrat, Neotoma fuscipes, in northeastern California. J. Mamm., 50:43-48.

MYKYTOWYCZ, R.

1962 Territorial function of chin gland secretion in the rabbit, Oryctolagus cuniculus (L.). Nature (London), 193:799.

1965 Further observations on the territorial function and histology of the submandibular cutaneous (chin) glands in the rabbit, Oryctolagus cuniculus (L.). Anim. Behav., 13:400-412.

NEE, J. A.

1967 The biology of the yellow-bellied marmot (Marmota flaviventris) in Sagehen Creek Basin. M.A. thesis, Univ. California, Berkeley.

1969 Reproduction in a population of yellow-bellied marmots (Marmota flaviventris). J. Mamm., 50:756-765.

NELSON, B. C., and C. R. SMITH

1976 Ecological effects of a plague epizootic on the activities of rodents inhabiting caves at Lava Beds National Monument, California. J. Med. Ent., 13:51-61.

NEWCOMBE, C. L.

1930 An ecological study of the Allegheny cliff rat (Neotoma pennsylvanica Stone). J. Mamm., 11:204-211.

NOIROT, E.

1972 Ultrasound and maternal behavior in small rodents. Dev. Psychobiol., 5:371-387.

OLSEN, R. W.

1973 Shelter site selection in the white-throated woodrat, Neotoma albigula. J. Mamm., 54:594-610.

ORIANS, G. H.
1969 On the evolution of mating systems in birds and mammals. Amer. Nat., 103:589-603.

PEARSON, P. G.
1952 Observations concerning the life history and ecology of the woodrat, *Neotoma floridana floridana* (Ord). J. Mamm., 33:459-463.

POLLITZER, R.
1954 Plague. W.H.O. Monogr. Ser., no. 22. Geneva: World Health Organization.

POOLE, E. L.
1936 Notes on the young of the Allegheny wood rat. J. Mamm., 17:22-26.
1940 A life history sketch of the Allegheny woodrat. J. Mamm., 21:249-270.

QUAY, W. G.
1953 Seasonal and sexual differences in the dorsal skin gland of the kangaroo rat (*Dipodomys*). J. Mamm., 34:1-14.

RAINEY, D. G.
1956 Eastern woodrat, Neotoma floridana: life history and ecology. Univ. Kansas Publ., Mus. Nat. Hist., 8:535-646.

RALLS, K.
1971 Mammalian scent marking. Science, 171:443-449.

REICHART, H. A.
1972 Small mammal populations in six Sagehen Creek basin habitats. M.S. thesis, Univ. California, Berkeley.

REYNOLDS, J. G.
1966 The midventral skin gland in three species of woodrat (*Neotoma*). M.A. thesis, Univ. California, Berkeley.

RICHARDSON, W. B.
1943 Woodrats (*Neotoma albigula*): their growth and development. J. Mamm., 24:130-143.

RICKLEFS, R. E.
1967 A graphical method of fitting equations to growth curves. Ecology, 48:978-983.

SALES, G. D., and J. D. PYE
1974 Ultrasonic communication by animals. London: Chapman and Hall.

SCHMIDT-NIELSON, K. S.
1964 Desert animals; physiological problems of heat and water. Oxford: Clarendon Press.

SHIRLEY, E. K., and K. S. SCHMIDT-NIELSON
1967 Oxalate metabolism in the pack rat, sand rat, hamster, and white rat. J. Nutr., 91:496-502.

SMITH, M. H.
1965 Dispersal capacity of the dusky-footed woodrat, Neotoma fuscipes. Amer. Midl. Nat., 74:457-463.

STONES, R. C., and C. L. HAYWARD
1968 Natural history of the desert woodrat Neotoma lepida. Amer. Midl. Nat., 80:458-476.

STORER, T. I., and R. L. USINGER
1963 Sierra Nevada natural history. Berkeley and Los Angeles: Univ. Calif. Press.

STRAUSS, J. S., and F. J. EBLING
1970 Control and function of skin glands in mammals. Mem. Soc. Endocrinol. (Hormones and the environment), 18:341-371.

SUMNER, L., and J. S. DIXON
1953 Birds and mammals of the Sierra Nevada with records from Sequoia and Kings Canyon National Parks. Berkeley and Los Angeles: Univ. Calif. Press.

THACKER, E. J., and C. S. BRANDT
1955 Coprophagy in the rabbit. J. Nutr., 55:375-385.

TURKOWSKI, F. J., and R. K. WATKINS
1976 White-throated woodrat (Neotoma albigula) habitat relations in modified pinyon-juniper woodland of southwestern New Mexico. J. Mamm., 57:586-591.

VAN VALEN, L.
1971 Group selection and the evolution of dispersal. Evolution, 25:591-598.

VAUGHAN, T. A.
1972 Mammalogy. Philadelphia: W. B. Saunders.

VERTS, B. J.
1963 Equipment and techniques for radio-tracking striped skunks. J. Wildl. Mgmt., 27:325-339.

VORHIES, C. T., and W. P. TAYLOR
1940 Life history and ecology of the white-throated wood rat, Neotoma albigula albigula Hartley, in relation to grazing in Arizona. Univ. Ariz. Coll. Agric., Agric. Expt. Sta., Tech. Bull., 86:455-529.

WALLEN K.
1977 Social organization in the dusky-footed woodrat (Neotoma fuscipes): field studies and laboratory experiments. Ph.D. diss., Univ. California, Berkeley.

WELLS, P. V., and R. BERGER
1967 Late Pleistocene history of coniferous woodland in the Mohave Desert. Science, 155:1640-1647.

WELLS, P. V., and C. D. JORGENSEN
1964 Pleistocene wood rat middens and climatic change in the Mohave Desert: a record of juniper woodlands. Science, 143:1171-1174.

WHIPPLE, I. L.
1904 The ventral surface of the mammalian chiridium. Zeit. Morph. Anthropol., 7:261-368.

WILSON, E. O.
1975 Sociobiology. Cambridge, Mass.: Belknap/Harvard Univ. Press.

WILSON, S. C., and D. G. KLEIMAN
1974 Eliciting play: a comparative study (Octodon, Octodontomys, Pediolagus, Phoca, Choeropsis, Ailuropoda). Amer. Zool., 14:341-370.

WOOD, F. DONAT
1935 Notes on the breeding behavior and fertility of Neotoma fuscipes macrotis in captivity. J. Mamm., 16:105-109.

Plates

PLATE 1. a. Typical den of Neotoma cinerea in rocks. Small accumulation of sticks is commonly the only external clue to nest site. White urine deposits are on rocks above den (Peaks site, Sagehen Creek). b. House of Neotoma fuscipes (Modoc County, Calif.), approximately 1.5 m high. Large accumulation of sticks and other objects is typical of many woodrat species.

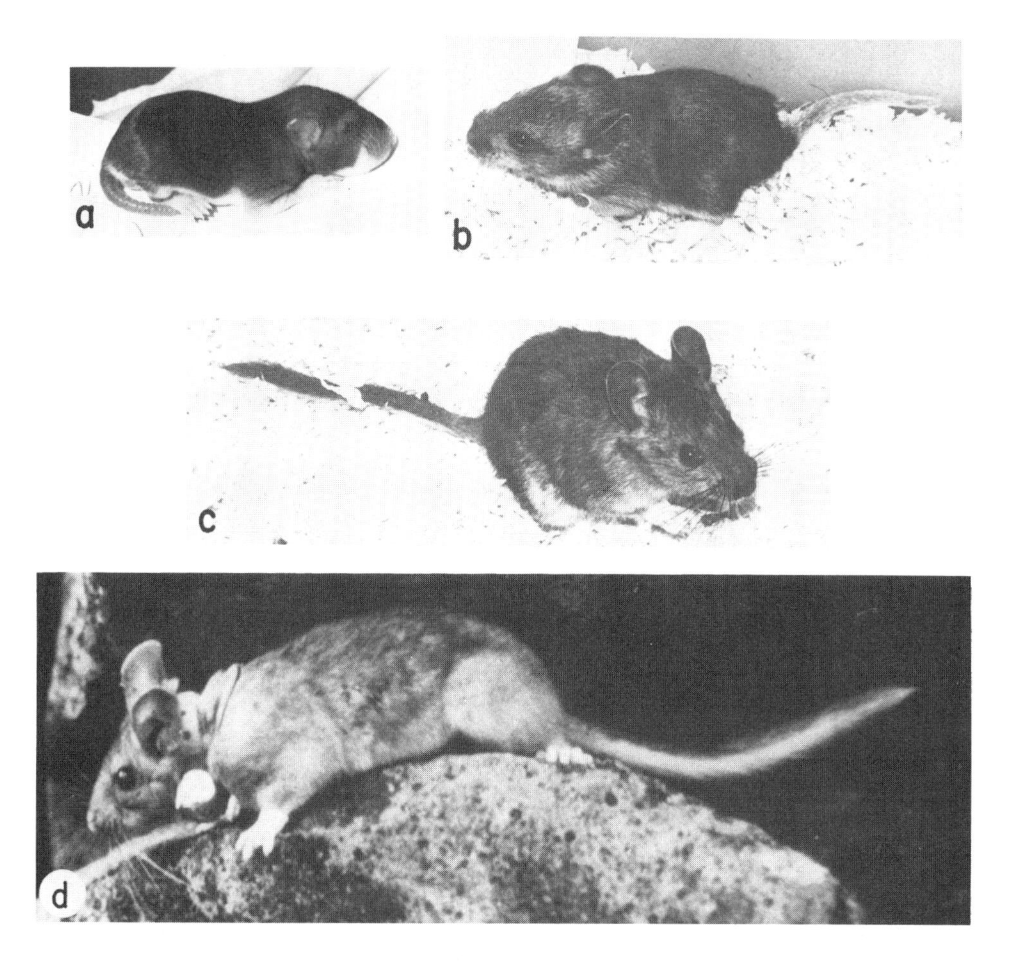

PLATE 2. a. Juvenile Neotoma cinerea, 4 days old. b. Early immature N. cinerea, 21 days old. c. Older immature N. cinerea, 50 days old. d. Subadult N. cinerea, about 3 months old. (Female 1173, wearing radio collar; photograph by J. G. Hall.)

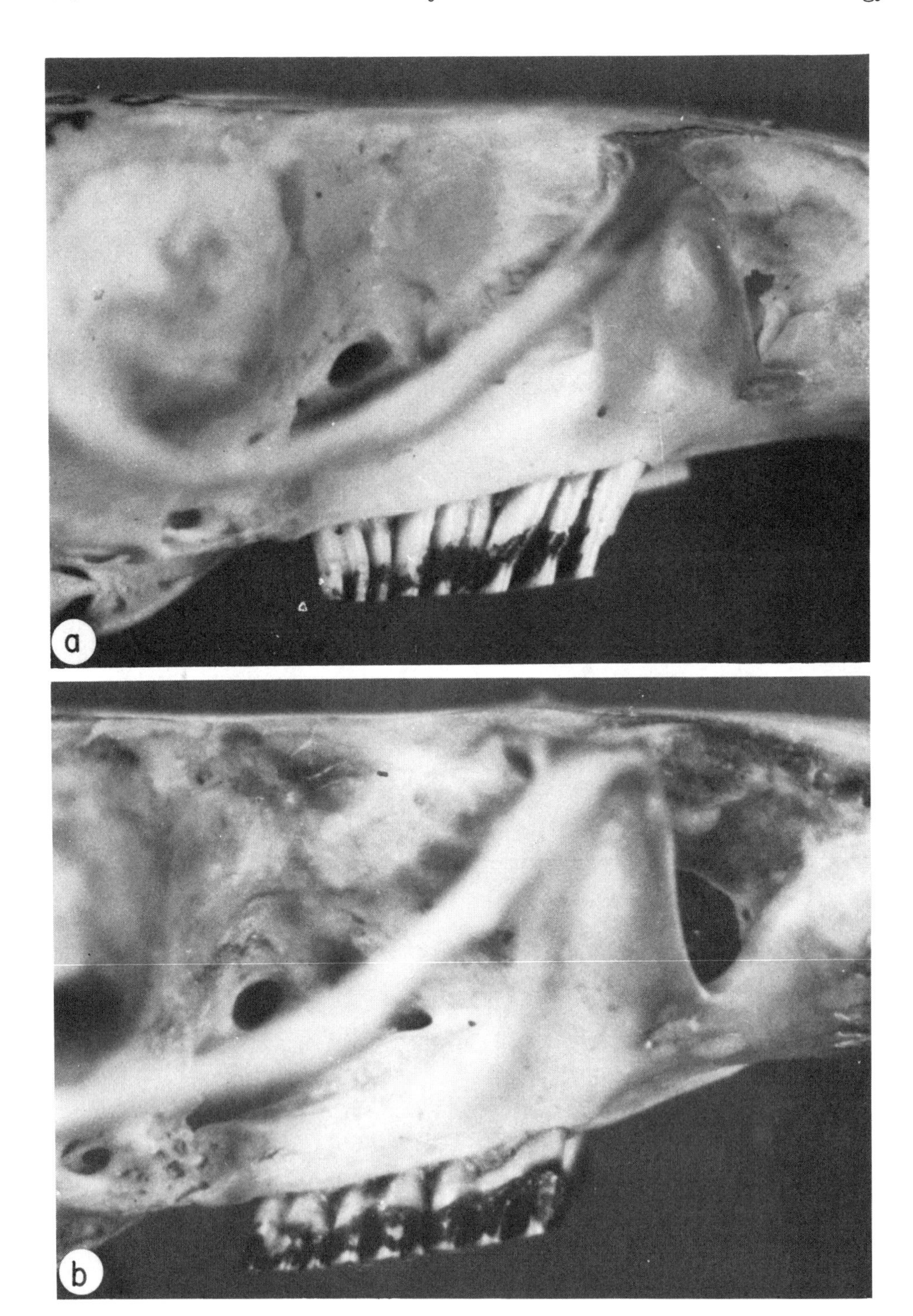

PLATE 3. a. Skull of subadult female Neotoma cinerea. (MVZ 72060, collected July 15, 1936, 3 mi W Swan Valley, Bonneville County, Idaho.) Folds of molars extend into alveoli; roots are not visible. b. Skull of adult female N. cinerea. (MVZ 126809, collected Oct. 8, 1960, at Sagehen Creek.) Folds are only on distal half of molars; roots are visible.

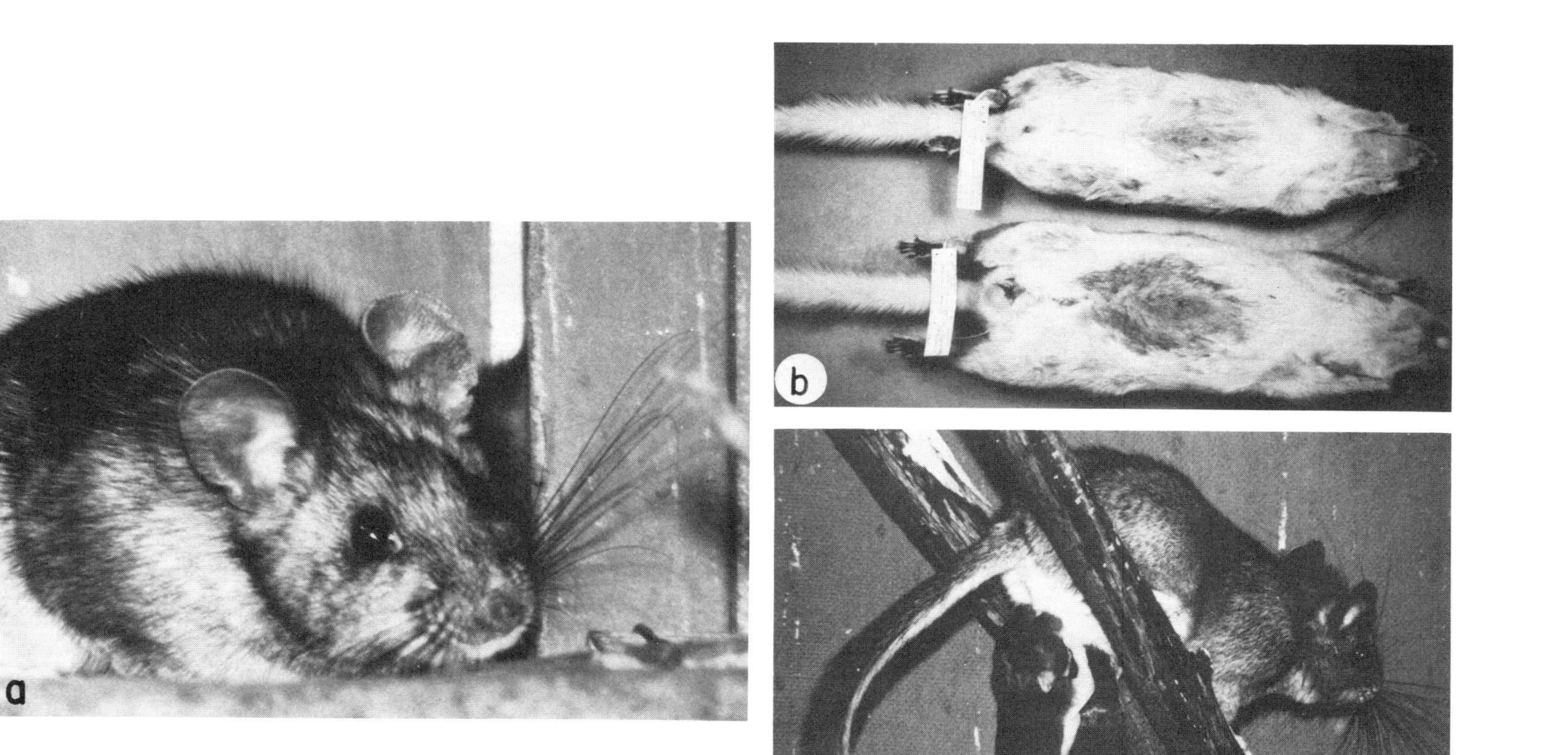

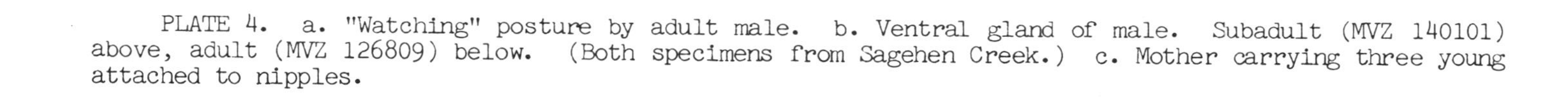

PLATE 4. a. "Watching" posture by adult male. b. Ventral gland of male. Subadult (MVZ 140101) above, adult (MVZ 126809) below. (Both specimens from Sagehen Creek.) c. Mother carrying three young attached to nipples.

PLATE 5. a. Rock outcrop (about 7 m high) with extensive urine deposits from N. cinerea (white edging along horizontal ledges). (1 mile N Susanville, Lassen County, Calif.) b. Close-up of urine deposit with small amount of fresh urine (arrows), indicating presence of one or few individuals. Fresh urine here was bright yellow, affording more contrast than can be seen in black and white. (Pen on right indicates scale.) c. Close-up of urine deposit with extensive fresh urine, indicating an especially active settlement with more than a dozen individuals. Fresh brown urine (color probably depends on diet) nearly covers entire white deposit.

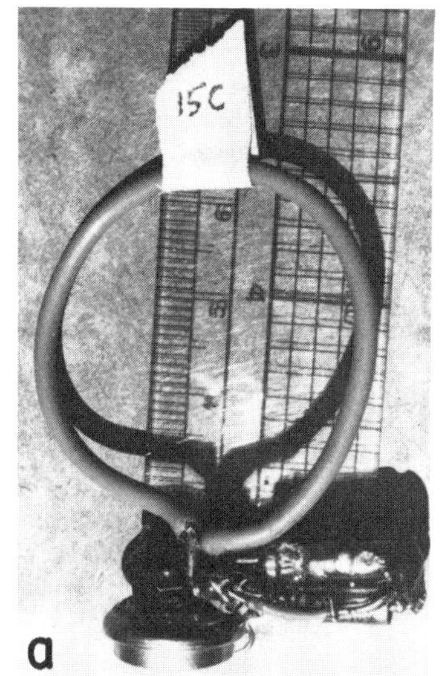

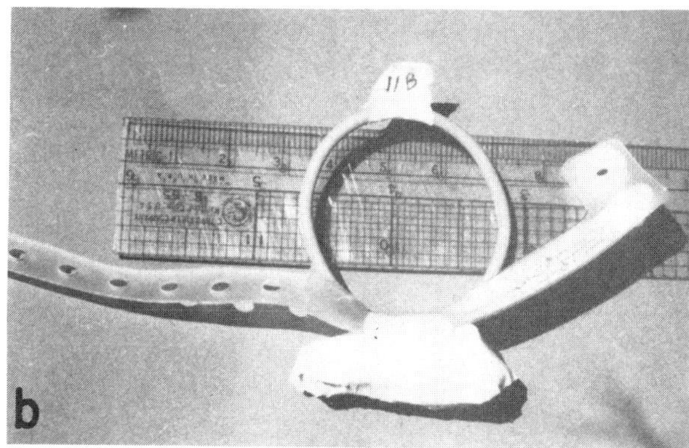

PLATE 6. a. Radio transmitter fully assembled, prior to wrapping and attachment of collar. b. Transmitter ready to attach to wcodrat. c. Woodrat (about 3 months old) wearing transmitter.

PLATE 7. a, b. Upper Spring study site. (See fig. 8 for camera positions.) c. Lower Spring study site. (See fig. 9 for camera position.)

PLATE 8. a. Rockslide study site. (See fig. 14 for camera position.) b. X89 study site, general view. Main denning area is in left center of view. Highest rock outcrops are out of sight above curve of hill. (See fig. 15 for camera position.) c. X89 study site, close-up of central den area (fig. 16). Circles indicate location of dens; numbers refer to individual residents as in fig. 18.

PLATE 9. a. Mile 4.2 study site. b. Example of more continuous denning habitat than found at Sagehen. (8.5 mi E of Likely, Modoc County, Calif.)